"I find this work by well-qualified, faith-filled scientists convincing – a book desperately needed in a day when critics consider the Scriptures and science to be in serious conflict and various well-meaning Christians fail to fully account for the available geological evidence."

— PAUL COPAN, PLEDGER FAMILY CHAIR
OF PHILOSOPHY AND ETHICS,
PALM BEACH ATLANTIC UNIVERSITY

"Here is a resource that not only demonstrates the superiority of an ancient-earth interpretation of the Grand Canyon but also testifies of God's grandeur and providence. This is a necessary textbook for any class on Creation."

— KENNETH KEATHLEY, PROFESSOR OF
THEOLOGY, SOUTHEASTERN BAPTIST
THEOLOGICAL SEMINARY

"The various authors of this book have done us all a tremendous service in their patient and clear exposition of geological thinking about the Grand Canyon (a magnificent place in its own right!)… These are God's processes after all! I urge everyone to read this, believer or not – you will enjoy it."

— C. JOHN COLLINS, PROFESSOR OF OLD
TESTAMENT, COVENANT THEOLOGICAL
SEMINARY

"Can Bible-believing Christians also believe that the earth is billions of years old and that the Grand Canyon could not have been formed by Noah's Flood?… On page after page, professional geologists explain that "flood geology"… fails to explain massive amounts of evidence in the Grand Canyon itself."

— WAYNE GRUDEM, RESEARCH PROFESSOR
OF THEOLOGY AND BIBLICAL STUDIES,
PHOENIX SEMINARY

"*The Grand Canyon, Monument to an Ancient Earth* allows this marvelous landform to tell its own story, walking the reader through an illustrated guide to the geologic trail exposed in the canyon walls. A must read for anyone interested in the direct testimony of Creation itself on the history of our world."

— DAVID R. MONTGOMERY, AUTHOR OF *THE ROCKS
DON'T LIE: A GEOLOGIST INVESTIGATES NOAH'S FLOOD*

"[T]he authors, most of whom are Christians, have systematically contrasted the claims of so-called 'flood geologists' with the actual details of Grand Canyon geology. In doing so, they have utterly demolished the flood hypothesis with an avalanche of geologic evidence sufficient to fill a canyon."

— DAVIS A. YOUNG, PROFESSOR EMERITUS OF
GEOLOGY, CALVIN COLLEGE

"The authors carefully, professionally, and convincingly show that flood geology is simply not science. They pose the question, 'Does it really matter?', and end with what we unquestionably should all agree with, 'Truth always matters'."

— JOHN W. GEISSMAN, PAST PRESIDENT,
GEOLOGICAL SOCIETY OF AMERICA

"*The Grand Canyon* … is terrific. It is a comprehensive primer on geology as we know and practice it… Everyone on Earth should be aware of what this book contains as a science text, as well as a rationale for employing scientific methods to a basic understanding of anything natural."

— JOHN WARME, PROFESSOR EMERITUS,
COLORADO SCHOOL OF MINES

(Complete endorsements are on pages 236-239 for Collins, Copan, Geissman, Grudem, Warme, and Young.)

"The one who states his case first seems right, until the other comes and examines him." PROVERBS 18:17

Alarcon Terrace and the Great Bend. *Photo by David Edwards.*

The *Grand Canyon*

MONUMENT
TO AN
ANCIENT EARTH

Can Noah's Flood Explain the Grand Canyon?

Gregg Davidson, Joel Duff, David Elliott, Tim Helble,

Carol Hill, Stephen Moshier, Wayne Ranney, Ralph Stearley,

Bryan Tapp, Roger Wiens, & Ken Wolgemuth

The Grand Canyon, Monument to an Ancient Earth:
Can Noah's Flood Explain the Grand Canyon?

Published by Kregel Publications, a division of Kregel, Inc., 2450 Oak Industrial Dr. NE, Grand Rapids, MI 49505.

Cover photograph: Wayne Ranney
Back cover photograph: Bronze Black

Senior editor: Carol Hill
Text editor: Gregg Davidson
Photo/Illustration editors: Tim Helble and Wayne Ranney
Copy editor: Anne Thomas
Book design and layout: Bronze Black Design
Kregel layout & editing: Susan Coman and Rachel Warren
Illustrators: Tim Helble, Rusty Johnson, and Paul Mitchell

Authors: Gregg Davidson, Joel Duff, David Elliott, Tim Helble, Carol Hill, Stephen Moshier, Wayne Ranney, Ralph Stearley, Bryan Tapp, Roger Wiens, & Ken Wolgemuth

ISBN: 978-0-8254-4421-0

Printed in the United States of America

1 2 3 4 5 / 30 29 28 27 26 25 24 23 22 21 20 19 18 17 16

Tulsa, Oklahoma
918-852-3082
www.solidrocklectures.org

Acknowledgments

It is a daunting task to adequately acknowledge the individuals and organizations to whom we are indebted for instructing, challenging, inspiring, and contributing to the work you now hold in your hands. Each of the eleven authors owes much to the teachers who passed on their knowledge, their love for their disciplines, and their enthusiasm for solving puzzles. Space does not allow for a list of all their names. We are especially grateful for the individuals who walked before us, on whose shoulders we stand – men such as Larry Kulp, Clarence Menninga, Howard van Till, Davis Young, and Daniel Wonderly – who embraced their faith and the study of the creation to challenge the unsubstantiated biblical and scientific claims of early proponents of Young Earth Creationism. (We feel greatly honored to have words of praise from Dr. Young among the endorsements found inside the cover!)

None of the authors are first time writers. All have experience in penning words, but few were familiar with the rigors of producing a book filled with high-quality photographs and original artwork. We are grateful to those who prompted us to begin, encouraged us to keep making progress when obstacles were encountered, and celebrated with us when the objective was realized. This appreciation goes first to our spouses, Alan Hill, Kristin Davidson, Helen Ranney, Tina Helble, Dawn Duff, Carol Moshier, Gloria Stearley, Gayle Tapp, Gwen Wiens, and Helen Wolgemuth, many of whom served as first readers of draft chapters. The book has benefited greatly from the insightful reviews of Greg Bennett, Scott and Grace Buchanan, Carol Christopher, Lorence G. Collins, Kenneth J. Van Dellen, Hillary Morgan Ferrer, Jerrie Hall, Jordan Howard, Robert W. Scott, David Vaughnn, and John Warme. Student reviews from Dr. Moshier's 2015 Geology of the National Parks class at Wheaton College added additional helpful insights. Rusty Johnson and Paul Mitchell (along with author Tim Helble) were long-suffering in responding to repeated requests to tweak illustrations, as were Bronze Black and Susan Coman with myriad corrections to the layout. Anne Thomas provided early copy editing and production guidance. Nicole Heidebrecht served greatly in organizing permissions and entering innumerable last-minute corrections to text, references, and the index.

We are particularly appreciative of the enthusiastic support and monetary contributions from the American Scientific Affiliation, the John Templeton Foundation, and the BioLogos Foundation. Individuals from these organizations that served as cheerleaders and advocates for many years include Randy Isaac, Paul Wasson, Andrew Rick-Miller, Darrel Falk, Deborah Haarsma, and Kathryn Applegate. Geoff Feiss and Kent Condie, from the Geological Society of America, also lent their encouragement in pressing for publication. The many professional photographs, original artwork, and professional layout for the book were made possible by donations from John Barrett, Edward Beaumont, Randall Cade, Benjamin Chenoweth, Lorence G. Collins, David Curtiss, Richard Daake, Marlan Downey, Dan Heinze, Michael and Karen Kuykendall, Robert W. Scott, M. Ray Thomasson, and others who remained anonymous.

Finally, we are indebted to Dennis Hillman and the staff at Kregel for their enthusiasm and willingness to take on a project they know not everyone will like. Their courage and commitment to providing people with the information needed to make informed decisions is greatly appreciated.

The Authors

CONTENTS

FOREWORD

by **Wayne Ranney**

The Grand Canyon of the Colorado River is one of our planet's most compelling and recognizable landforms. Each year, nearly 4.5 million people stand in awe on the canyon's rim and gaze into one of the most colorful and spectacular panoramas that can be viewed from a single place (Fig 1). The canyon overwhelms the senses and inspires deep feelings within those who look into it. It is the rare individual who, upon seeing the Grand Canyon for the first time, does not ask, "How could this have formed?"

Humans have interacted with this deep gorge since prehistoric times, but we do not know how ancient people reacted upon first seeing the canyon's vast excavated space. We know they were here because of the cryptic evidence left behind in caves tucked into the canyon walls. Like us, they too must have been awed.

Old World Europeans did not arrive at the canyon until the year 1540, when Hopi Indian guides led a group of Spanish gold seekers from the Coronado Expedition to its South Rim. Because the canyon did not contain gold or other precious metals, the Spaniards moved on and the area lay mostly unvisited by European immigrants for more than three centuries. Not until the arrival of the first explorers in the mid-nineteenth century did the Grand Canyon penetrate the American consciousness, by way of various reports sent to Congress by the expedition leaders or the scientists who accompanied them.

In the exhilarating days of Western exploration, much of the American West was called the "Great American Desert." At that time, visions of earthly beauty were based exclusively upon the European ideal, honed in the Alps and rich with idyllic scenes of

Figure 1. The heart of the Grand Canyon from Mather Point. *Photo by Mike Buchheit.*

Figure 2. The steamship Explorer, the craft used by the Ives Expedition to travel up the Colorado River in the winter of 1857. *Courtesy of the Library of Congress.*

Figure 3. John Strong Newberry, first geologist to see the Grand Canyon. *Photo from the U.S. Geological Survey.*

glaciated mountains, clear-running streams, and the color green. Americans were largely repulsed by the vast, arid Western landscape and observed it only in passing as they crept along in their creaking wagons towards greener pastures in Oregon and California.

Ultimately, however, Manifest Destiny brought them face to face with the southern edge of the Colorado Plateau and the Grand Canyon, and it was not always an easy encounter. Hampered by the dry winter conditions in 1857 and by fleeting native guides who often abandoned the group without warning, the leader of the first American expedition to ever lay eyes on the canyon was not impressed (Fig 2). Hailing from the farms and hardwood forests of his native New Hampshire, Lt. Joseph Christmas Ives called the Grand Canyon a "profitless locality" and decreed that it would be "forever unvisited and undisturbed." He was wrong. Had he taken the time to ask his colleague, John Strong Newberry (a member of his own expedition, no less), Ives would have found a much different response to the "Big Cañon," as it was called at that time (Fig 3).

Newberry accompanied the Ives Expedition as their geologist, and was likely the first Anglo to see the Grand Canyon and appreciate it as something special and unique on our planet. About the Colorado Plateau he wrote, *"[These canyons] belong to a vast system of erosion, and are wholly due to the action of water. Probably nowhere in the world has the action of [water] produced results so surprising."*

He was spot on. Having the trained eye of a scientist, Newberry recognized that the Grand Canyon was not a giant earth fissure that only later became occupied by the Colorado River, but rather that the giant canyon was actually made by the river. Newberry and subsequent American geologists would use their experience here at the Grand Canyon and in the American West to further develop a branch of geology known as *fluvialism,* which investigates how the Earth's surface can be shaped by the action of rivers. Geologists in

Figure 4. Iceberg Canyon, Colorado River, Wheeler Expedition. *Photo by T. H. O'Sullivan, 1871. Courtesy of the Library of Congress.*

Europe had developed the concept of fluvialism early in the nineteenth century, but never imagined the magnitude of a river's cutting power before the deeply carved landscapes of the American Southwest were discovered, especially the Grand Canyon.

Thousands of geologists have followed in Newberry's footsteps, with George Wheeler (Fig 4), John Wesley Powell, G. K. Gilbert, Clarence Dutton, and Charles Walcott being only a notable few of his nineteenth century contempo-

Figure 5. John Wesley Powell as he looked in the 1870s when he completed his second expedition down the Colorado River.
Photo from the U.S. Geological Survey.

raries (Fig 5). All of them were greatly impressed by the Grand Canyon's colorful rock layers, extreme topography, and enigmatic origin (Figs 6, 7). Just as the archaeological evidence has given us clues about the former presence of our prehistoric forebears, so too do the rocks here reveal evidence of their origin. It took centuries, even millennia, for humans to comprehend that the history of our planet is contained in rocks, but once this insight was achieved, the story of planet Earth began to be understood.

Exploring and discovering the Earth's geologic story was not about challenging religious beliefs. The earliest scientists in the days after the Reformation believed that science could show the precise manner in which the Creator had accomplished the task of creating natural phenomena such as the Grand Canyon. As more and more evidence accumulated, scientists, many of whom were outspoken Christians, became increasingly convinced that the history of the Earth was vast and complex.

Somewhat surprisingly, it was not until the beginning of the twentieth century that religious opposition began to be voiced against the antiquity of the Earth. In the 1920s, the Scopes Trial brought that controversy to the forefront of public attention. Thus, we find modern society confused and polarized on the topic.

Some who seek to discredit science nevertheless use modern devices bequeathed to them through scientific discovery and experimentation (such as medical advances), while they simultaneously discount other aspects of science that seem to conflict with their interpretation of the Bible.

To some people, the pronouncement that our planet was created "billions of years ago" is viewed as an attack on the Bible, but it need not be. It is a little-known fact that among modern professional geologists today who embrace the inspiration and authority of the Bible, the vast majority also understand the Earth to be billions of years old. These and other scientists observe and report their findings, much as Galileo, Newton, and others have done before them, seeking to find the most plausible explanations for what they observe. It is to be hoped that this controversy will, over many years, end much as the one with Galileo did.

Figure 6. Powell's boat trip down the Colorado River of the Grand Canyon in the 1870s. Hand-colored woodcut.
North Wind Picture archives, EXPL2A-00144.

In this book, you will find answers to questions you may have about the science of geology and how it works to arrive at certain conclusions. Geologists use many of the same scientific methods and techniques that have given us our televisions, microwave ovens, and cell phones. The petroleum that powers our automobiles has historically been discovered using methods arrived at wholly from an old earth viewpoint.

Flood geology, the idea that Noah's flood was responsible for laying down most of the sedimentary rock record, does not provide a credible scientific understanding of how oil and gas reservoirs have formed, with the trillions of barrels of oil, and enormous quantities of gas that fuel our modern society. Though few people realize it, to deny an old age for the Earth or the Grand Canyon, while embracing other aspects of science, is essentially a statement that science works only when we agree with the outcome. In this book, you will find explanations of how the Grand Canyon came to look the way it does, along with assessments of Young Earth/flood geology arguments to the contrary.

Many of the contributors to these chapters are Christians, while some are not. However, each of us is a student of the Earth who is troubled by what we believe to be a needless controversy that surrounds the story of the Grand Canyon. None of us presumes that acceptance of great age for the Grand Canyon will undermine religious faith. In fact, one early chapter offers insights into why an old earth view is actually *more* in line with biblical teaching.

In 150 years of scientific study, we have learned a great deal about how the Grand Canyon was formed. But like other branches of science, the investigative work continues and the story is not yet complete. This is because the Colorado River is an agent of erosion, and as it has become deeper and wider through millions of years, it has removed some of the evidence for its origin. But on the other hand, much *is* known about the formation of the canyon, and that is the story we present in this book.

It is truly remarkable that we humans, though tiny and short-lived, can yet peer deep into the past and reconstruct events never witnessed by human eyes. The Grand Canyon, by virtue of its grand exposures, has enabled us to not only be awed by its visual grandeur, but also to be enthralled by the incredible depth of time and richness of its history. Perhaps the greater lesson gleaned from the scientific study of the Grand Canyon is the humbling recognition that our own existence and inquiries, significant though they be, represent but a tiny sliver of the Earth's story.

Figure 7. The grandeur of the Grand Canyon from Point Sublime; colored woodcut by William H. Holmes, 1882. *Courtesy of the Library of Congress.*

View from Mather Point. *Photo by Tom Bean.*

PART 1
Two Views

Central Grand Canyon. *Photo by Mike Buchheit.*

CHAPTER 1

INTRODUCTION

by **the Authors**

THE GRAND CANYON IS A LAND OF superlatives! It is no wonder that when most people think of "Earth history," images of the canyon come to mind. It is a mile-deep chasm that the Colorado River and its tributaries are carving deeply into the southwestern edge of the Colorado Plateau – a geographic province that holds other landscape wonders, such as the Petrified Forest, Monument Valley, and the Canyonlands. The Grand Canyon extends for 277 miles, from the Utah-Arizona border at Lees Ferry to the Arizona-Nevada border and the upper reaches of Lake Mead (Fig 1-1). For those who explore the river in rafts, locations are identified by the river miles, beginning with River Mile 0 at Lees Ferry and ending at River Mile 277 at the mouth of the canyon at Grand Wash Cliffs (Figs 1-2, 1-3, 1-4).

Along this stupendous river corridor and extending far up into many of the side canyons is Grand Canyon National Park, encompassing over 1.2 million acres of splendid geology. Grand Canyon Village, located on the South Rim, is nearly 7,000 feet above sea level, while the visitor area on the North Rim is 8,200 feet above sea level, and Phantom Ranch, down at the river, is 2,460 feet above sea level. The maximum depth of the Grand Canyon is over 6,000 feet, and its width varies from less than one mile across in Marble Canyon to 18 miles across

at points downstream (Figs 1-1, 1-5). The deepest parts of the canyon contain three sections with inner gorges – deep, narrow, steep-walled canyons cut into crystalline rock (Fig 1-7). These crystalline rocks are overlain by 4,000 to 5,000 feet of sedimentary rock exposed in the canyon. Thousands of feet of additional rock layers overlie these layers to the north of the Grand Canyon. The immensity of the chasm, both in its depth and breadth, creates an incomparable panorama of cliffs, amphitheaters, mesas, buttes, and temples.

Two Different Views of the Grand Canyon

The modern (or "conventional") geologic understanding of the origin of the Grand Canyon recognizes eons of major land changes, including rising and falling sea levels, land uplift and subsidence, long periods of deposition and erosion, faulting and folding of rock, and the eventual carving of most of the canyon by the Colorado River and its tributaries.

In opposition to this understanding are some Young Earth Creationists who have promoted a view that makes two bold claims: (1) that a biblical worldview leads necessarily to the conclusion that the Grand Canyon and its rocks were created

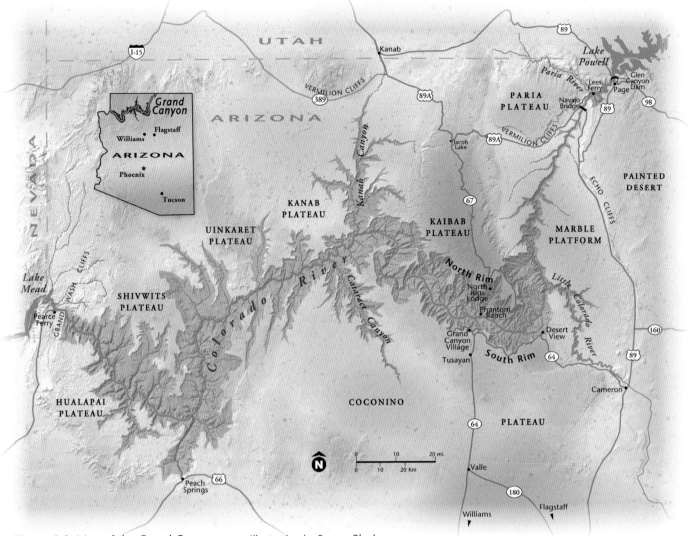

Figure 1-1. Map of the Grand Canyon area. *Illustration by Bronze Black.*

in recent events associated with Noah's flood, and (2) that the scientific evidence, when not bound to an evolutionary or uniformitarian bias, speaks clearly of a recent global deluge. Those seeking to use science to defend this view are generally referred to as *flood geologists.*

Somewhat ironically, both the conventional geologic view and the Young Earth view concur that the Grand Canyon's layers and deep chasm formed by *natural* processes, and are subject to scientific inquiry. However, they differ radically in the time it took to form the canyon and in the approach used to investigate and interpret the evidence. The purpose of this book is to show that the first Young Earth claim is not supported by a straightforward reading of Genesis, and that the second claim is not supported by an equally straightforward study of the geology of the canyon.

For any view of the history of the Grand Canyon to be considered plausible, it must be able to provide consistent explanations for the origin of not just isolated layers or features, but for the entire sequence of layers, along with all the internal chemistry, structures, and fossils. We'll begin our exploration of the canyon with a simple overview of what is visible from a walk along the South Rim.

A Walk Along the Rim

Let's enter Grand Canyon National Park from its east side, where Highway 64 skirts the South Rim (map above). There are several designated overlooks along this route, including Desert View, with its historic Watchtower, and Lipan Point, which give us a magnificent view of some of the canyon's deepest layers (Figs 1-6, 1-8). When peer-

Figure 1-2. Lees Ferry at River Mile 0; viewed from Vermilion Cliffs, looking south downstream towards Marble Canyon. *Photo by Tim Helble.*

ing over the edge at Lipan Point (Fig 1-8 on page 19), what immediately captures your attention? No doubt the immensity, the incredible depth, and the breathtaking beauty of the canyon — but after that, what strikes you about the rock layers of the canyon?

The most obvious observation is that the rock layers are not all of the same type or orientation. The upper series of rock layers are horizontal, while the lower layers are tilted down towards the east. You can also see distinctly different coloration

Figure 1-3. Raft trip starting from Lees Ferry. *Photo by Tim Helble.*

Figure 1-4. End of the Grand Canyon, where the Colorado River flows into Lake Mead, near River Mile 277. *Photo by Tim Helble.*

Figure 1-5. The Little Colorado River (lower right, robin-egg blue) enters the Colorado River below Cape Solitude, River Mile 62. Photo is looking north up Marble Canyon, opposite the direction of Figure 1-2. *Photo by Tom Bean.*

— some gray, some reddish, and some buff-colored. Some layers are thin, some are thick. Some layers form cliffs and look like they might be composed of harder rock; others look soft and weathered, and form gentle slopes covered by debris from above. If you have a good eye for elevation differences, you may also notice that the layers on the opposite, North Rim, side of the canyon are significantly higher than the layers beneath your feet.

Recognizing that the entire Grand Canyon is not visible from one spot, we'll travel westward, towards Grand Canyon Village, until we get to Yavapai Point. Here we have another look over the edge (Fig 1-9 on facing page). The bottom of the canyon looks quite different here than it did along the Desert View rim. The tilted beds and wide inner canyon are gone, replaced by a deep, narrow gorge in crystalline rocks that have no visible layering at all. Where did the tilted layers

Figure 1-6. Desert View Watchtower at the east entrance to the Grand Canyon. *Photo by Mike Buchheit.*

Figure 1-7. The Inner Gorge of the Grand Canyon, near River Mile 83. *Photo by David Edwards.*

go? Clearly, the history of this amazing landscape is complex! How did these layers get here, how were they deformed or altered over time, and how was the canyon carved?

Any explanation must account for not only individual features observed in the canyon, but also for how all the pieces fit together in a larger story. The chasm between the explanatory models of flood geology and conventional geology is as wide as the canyon itself. Does each fit the pieces of the story together equally well? Keep reading, and we'll see!

Figure 1-8. View from Lipan Point. *Photo by Mike Koopsen.*

Figure 1-9. Monsoon Storm seen from Yavapai Point. *Photo by Mike Koopsen.*

Drawing of a nearly completed Noah's Ark, from Johann Scheuchzer's *Physica Sacra* (1731).
Courtesy David Montgomery Collection.

CHAPTER 2

WHAT IS FLOOD GEOLOGY?

by **Stephen Moshier** and **Carol Hill**

T HE ACCOUNT OF NOAH'S FLOOD, AS described in Genesis, chapters six through nine, is woven into the fabric of three major faiths: Judaism, Christianity, and Islam. According to the Hebrew scriptures, God warned Noah of an impending flood and instructed him to build an ark (a large boat) to preserve his family and pairs of every animal (Fig 2-1). Rain fell for forty days and nights, and water rose up from the "fountains of the deep" over the next five months as Noah and his family rode out the storm. All of the other people and animals "on the face of the earth" died in the flood (Fig 2-2). Water began to recede after the fifth month, and after one year, the land was finally dry enough for Noah's family and the animals to exit the ark (Fig 2-3).

"Flood geology" derives from an interpretation of these scriptures leading to the belief that Noah's flood did not just cover the human population, but instead was global in extent and left a massive record that is preserved in the rocks, fossils, and landforms of planet Earth. Since the Grand Canyon displays one of the most extensive and dramatic exposures of sedimentary rocks on Earth, flood geologists have focused their attention on the canyon's layers as evidence of the biblical flood. Yet, nearly all modern geologists – including most Christian geologists – find no physical support for that interpretation of Grand Canyon geology.

Figure 2-1. The building of Noah's Ark. *Painting by Aureliano Milani, The Museum of Fine Arts, Budapest, Hungary.*

Figure 2-2. "The Deluge" by Francis Danby, 1840.
Tate Gallery.

In this section we present a brief history of the encounter between science and Christian theology, and we introduce the ideas of Young Earth Creationism and flood geology. We will review and critique the basic principles of flood geology and how Young Earth advocates extend their doctrinal views to the geology of the Grand Canyon.

Flood Geology and Modern Geology: Historically Intertwined

Modern science emerged in the Western world during the Renaissance and was largely influenced by the biblical view of an orderly, comprehensible creation. The science of astronomy, particularly after the invention of the telescope, led to the replacement of a long-held geocentric (Earth-centered) view of the solar system with a heliocentric (Sun-centered) view. Initial resistance from the church peaked at the time of Galileo's discoveries, but it soon faded as the evidence for a Sun-centered system grew and as people recognized that the particular scriptural passages in question were not inherently in conflict with Galileo's assertions. The age of creation was not yet a concern for scientists, who had no evidence or reason to question estimates that were based on biblical genealogies and the six-day creation account in Genesis 1. It is important to remember that early scientists, from the Renaissance to the nineteenth century, including

Galileo, typically sought to use their newly acquired skills of observation and scientific reasoning to show proof for how God created the Earth. They did not seek to overturn or overrule Scripture.

As Renaissance astronomers gazed skyward, other natural philosophers were beginning to look closer to home for signs of how the Earth had been formed. In Denmark, Nicholas Steno (1638–1686) was describing rock strata and the principles for interpreting them. Men like Steno assumed that catastrophes of biblical proportions were responsible for rock layers and the chiseled features of the landscape. The discovery of fossil fish and shells embedded in mountain rocks seemed proof positive of at least one global deluge. Even the great Leonardo da Vinci (1452–1519) was interested in the idea of a catastrophic deluge, and he devoted pages of his famous notebooks to sketches of torrents of water moving through cataracts in the mountains. However, da Vinci came to the independent conclusion that violent currents did not transport the fossil shells found in the mountain rocks high above his own village, because he observed that the shells appeared to be oriented in the rock just as shells accumulate today on the undisturbed seafloor.

Geology became a true scientific discipline near the beginning of the nineteenth century. Two schools of thought emerged among the first geologists. Catastrophists, or diluvialists (the original flood geologists), did not necessarily believe the Earth is young, but held to the view that the Earth's history was shaped by one or more violent episodes, with Noah's flood being perhaps the most recent. The other group, the uniformitarians, saw uniformity in natural laws and forces that allowed them to use observations of present natural processes and environments to identify similar processes and environments in ancient rocks. For example, da Vinci's shells in mountain rocks compare favorably with collections of shells along many contemporary seashores because they formed the same way.

Over years of observation, this view eventually led them to conclude that the Earth is very ancient, and its history far too complex to be explained by a single catastrophic event. Note that this was not a rejection of the Noachian flood, nor was it a rejec-

Figure 2-3. Noah's Ark resting in the Mountains of Ararat, from Johann Scheuchzer's *Physica Sacra* (1731). *Courtesy David Montgomery collection.*

possibility of an ancient creation and did not challenge the philosophical basis of uniformitarian geology.

Yet, one religious denomination closely associated with fundamentalist teaching of the Bible demanded that a recent creation and a global flood were implicit in Genesis 1. In the early twentieth century, Seventh-day Adventists affirmed both events on the basis of the prophecies of their leader, Ellen Gould White (1827–1915). As an apologist for Adventist theology, author George McCready Price (1870–1963) rejected the uniformitarian interpretations of geology that supported an ancient Earth. Instead, Price promoted flood geology, which was his reinterpretation of geologic strata to conform to a single global deluge. He sought, but did not receive, support of his views from Christian professional geologists of his day.

Flood geology eventually gained followers in other denominations, including Bible professor John Whitcomb (b. 1924) and engineering professor Henry Morris (1918–2006). Their co-authored book, *The Genesis Flood: The Biblical Record and Its Scientific Implications* (1961), expanded on Price's work (though giving minimal credit to Price) and introduced flood geology to a wider fundamentalist and evangelical Christian audience. Their book established the modern Young Earth Creationism movement as a major cultural and political force. In fact, many of the claims of flood geology, as applied to Young Earth books such as *Grand Canyon: Monument to Catastrophe* (1994), and *Grand Canyon: A Different View* (2003), had their origins in *The Genesis Flood*.

tion of God's providence in nature – it was simply a rejection of the flood as a major cause of Earth's many geologic layers.

By the beginning of the twentieth century, catastrophist geologists had largely conceded to the uniformitarians. Few readers may be aware that the widespread rejection of Noah's flood as a major geologic episode and the likely antiquity of creation beyond 6,000 years did not concern most conservative Bible scholars and theologians at that time. Evidence of this is found in the documents known as *The Fundamentals,* which were written by conservative Christian pastors and seminary professors (published between 1910 and 1915).

Regarded as the foundation of the Christian fundamentalist movement, essays written on the relationship of science and Christian faith acknowledged the

DO ALL CREATIONISTS BELIEVE IN A YOUNG EARTH?

For some, adding "Young Earth" in front of "Creationist" seems redundant, but there are many who believe that God created the universe (technically making them "creationists") over a long period of time, which is consistent with conventional geologic understanding. In this book, the challenge is not against creationist belief, but with the Young Earth claims regarding the timing and processes at work in shaping Earth history.

Thus, the perceived disconnect between modern geology and the Bible emerged essentially from the beliefs of a small group of textual literalists in the early twentieth century. When viewed in relation to the beliefs of many earlier Church theologians and the first Renaissance scientists (who set out to reveal God's glory), the common claim that all biblically minded people believe in a young Earth has little historical precedence.

Flood Geology: Basic Principles

The principles of flood geology, as well as of Young Earth Creationism, are based in a selectively literal reading of Genesis. Here we review some of the most important scriptural interpretations leading to this view of geology and Earth history.

Age of the Earth and Date of the Flood

The Earth was created approximately 6,000 years ago, according to a 24-hour-day/six-days-of-creation story (Genesis 1) and the chronologies of Genesis 5 and 11. The Genesis flood is understood to have happened about 4,500 years ago. One popular Young Earth organization claims the year of the flood was 2304 BC, stating that "all civilizations discovered by archaeology must fit into the last 4,285 years," without any further explanation.

Source of Floodwater

No rain fell on the "earth" (interpreted to be "planet Earth," rather than "ground") before Noah's flood (Genesis 2:5). Rather, a "mist" (Genesis 2:6) served to moisten the ground from creation to the time of Noah's flood. Some flood geologists, especially those in the middle to late twentieth century, proposed that the mist of Genesis 2:6 refers to a dense vapor canopy that shrouded the Earth before the time of Noah's flood. However, in recent years flood geologists have grown increasingly skeptical of this idea.

Genesis 7:11-12 states that the windows of heaven were opened and all the fountains of the great deep were burst forth. Some flood geologists interpret this verse to mean that all of the water in the vapor canopy fell as rain and that a great amount of water in the Earth's crust was expelled along faults and from volcanoes, while others emphasize seawater jetting skyward along large fractures in the Earth's crust. Rain fell for forty days and nights, water continued to rise and flooding prevailed for one hundred and fifty days, and the flood then receded over the next one hundred and fifty days.

Extent and Geologic Results of a Global Flood

The Bible says that "all the earth" was flooded, with even the mountains being covered to a depth of fifteen cubits (Genesis 7:19–20), and that "all flesh" died (Genesis 7:21). Therefore, flood geologists infer that Noah's flood must have left an immense record of itself in the form of sedimentary rock containing fossils. In addition to a worldwide deluge, Earth's tectonic forces were violently at work, causing great mountain ranges to heave upwards, which is why rocks containing fossils are found on the highest mountain peaks. In other places, violent volcanism rapidly built up entire mountain ranges, including the mountains of Ararat reaching nearly 17,000 feet, where the ark eventually landed (Fig 2-4, and box below).

Initially conjoined continents like South America and Africa separated swiftly, moving to near their present positions on the globe during the flood year

INCREDIBLE (NONBIBLICAL) EVENTS REQUIRED FOR MOUNT ARARAT

Mount Ararat is an ancient volcano, with fossil-bearing sedimentary rock on its flanks interspersed between layers of lava. Consider the sequence of events required by flood geology for this to form during a single flood: (1) sediments and dead animals were deposited from flood waters; (2) the sediments turned into fossil-rich rock; (3) magma was extruded into and up through the sedimentary rock (which was miraculously shielded from significant alteration by the heat) to create an entire volcanic mountain range up to 17,000 feet in height; (4) this huge, once-melted rock mass cooled at a miraculous rate; (5) Noah's Ark landed on it – *all in 150 days!*

Figure 2-4. Mount Ararat, with Khor Virap Monastery, Armenia in the foreground. *Photo by Andrew Behesnilian.*

(referred to as "catastrophic plate tectonics"). Since even the highest mountains were covered, many Young Earth advocates believe the ark landed on the tallest peak of the Middle East region, Mount Ararat (elevation 16,803 feet). After the ark landed on Mount Ararat (or more precisely, on the "mountains of Ararat;" Genesis 8:4), the floodwaters began to decrease rapidly because of evaporation (Genesis 8:1) and also because they "returned from off the earth continually" (Genesis 8:3) to low elevations relative to the mountains that were raised during the flood. About one year after the flood started, the post-flood landscape where Noah landed was dry (Genesis 8:14), and the topography of planet Earth was completely changed from its pre-flood landscape.

"The Fall" and Animal Death

Young Earth Creationists believe that a world of peaceful coexistence of immortal creatures prevailed on planet Earth before Adam and Eve sinned by eating the fruit of the Tree of the Knowledge of Good and Evil (Genesis 3:6). The declaration by God that his creation was "good" (Genesis 1:25, 31) is understood to imply that there was no death or decay anywhere on Earth at that time. This act of disobedience in the Garden of Eden, known as *the fall,* brought physical death upon Adam and Eve and all subsequent generations of every living creature. Thus, *all* sediments containing fossils (almost all sedimentary rocks) *must* date to sometime after the fall.

Flood Geology: Biblical Problems

Multiple *scientific* arguments can be made, and have been made for centuries, against a global interpretation of Noah's flood. These arguments are reflected in questions such as, "How could the ark carry all of the animal species on Earth?" or, "How did llamas from South America or penguins from Antarctica migrate to and from the ark?" Here, we present some *biblical* arguments *against* flood geology.

A Literal Interpretation of Scripture?

Young Earth writers affirm the inerrancy of Scripture in its original form as the "one basic premise" informing their understanding of creation history. For them, biblical inerrancy means that words in the Bible are taken literally with little or no regard for how those words may have held different meanings when they were written. Readers of Young Earth literature are warned that nonliteral interpretations of words and phrases such as "day" and "all the land" or "all flesh" are compromises to accommodate evolutionary ideas about creation that violate biblical admonitions such as Deuteronomy 4:2: *You shall not add to the Word which I am commanding you.*

However, flood geologists frequently bring modern scientific concepts to the biblical text in a way that does just that – adding to the Word. Consider how Psalm 104:8 is "quoted" in *Grand Canyon: A Different View*: "*The mountains rose, the valleys* [ocean basins] *sank down to the place which You established for them*" (brackets in original). Here, the author feels free to interpret "valleys" to mean "ocean basins" even though this is not a literal translation (and thus, contrary to the book's stated "one basic premise" of biblical literalism).

Scholars and leaders of evangelical Christianity, which includes Christian fundamentalism, have thoughtfully considered matters of interpreting the Bible in light of modern scientific knowledge of Earth and cosmic history. One effort to defend the authority and truthfulness of the Bible, *The Chicago Statement of Biblical Inerrancy* (1978), clearly disavows ironclad biblical literalism that disregards ancient cultural contexts, literary forms, and phenomenological language (words corresponding to the perception of the author/reader).

Flood geologists have largely ignored these admonitions and have drawn geological and paleontological conclusions about the extent of the Genesis flood from many words and verses, without considering the ancient cultural context of the Bible. In truth, the flood geology position derives not from a literal interpretation of Genesis, but from debatable assumptions about the intended meaning of specific words and phrases. Several examples follow.

Eretz

The Hebrew word for land, *eretz*, has been translated as "earth" in many English Bibles. Flood geologists understand *eretz* to mean the entire planet Earth. But, *eretz* literally means "earth, ground, land, dirt, soil, or country." For example, in the creation account (Genesis 1:10), "God called the dry land Earth (eretz)," which is clearly not a reference to the planet as a whole. References to the "face of the ground" (also eretz) in other verses are best understood as the *soil* or *local region* that we can see around us; that is, what is within view of the horizon. The ancient Hebrews knew nothing about planet Earth. The land around them was all they knew. *The mistranslation of eretz as planet Earth, instead of as a local parcel of land, is the foundation for many of the false assumptions and precepts of flood geology.*

"All," "Every," and "Under the Whole Heaven"

Flood geologists also emphasize the global extent of the flood by such expressions as "all," "every," "under the whole heaven," and "all the world," which occur throughout the Genesis account. Yet, there are numerous examples from the Old Testament (and other literature of its time, such as Akkadian texts) where the expressions of "all" or "every" could never have been understood to mean global. For example, when "*all the countries came*

THE PROBLEM OF OIL

The account of Noah in Genesis refers to bitumen (Genesis 6:14), a tarry substance found at oil seeps, which was used to waterproof the hull of the ark. Some Bible translations use the word "pitch" instead of bitumen, which could mean tar from tree sap, but the context for the Hebrew word that is used throughout Genesis is bitumen from oil seeps or tar pits. Oil derives from the remains of microscopic organisms, and is invariably associated with thick sequences of deeply buried rock. So, how did so many organisms live, die, get deeply buried, turn into bitumen, and seep to the surface in the few thousand years between the fall of Adam and Eve and the flood?

to Egypt to buy grain from Joseph, because the famine was severe in all the world (Genesis 41:57)," there is no reason to insist that human groups or tribes from Europe, Asia, the Americas, and Australia were included. Rather, the most straightforward reading of this passage would understand that from the author's perspective, all the countries of the then-known world came to Egypt. So why is a global interpretation insisted upon by flood geologists for Noah's flood?

"Covered the Mountains..."

The depth of the flood is also a matter of interpretation. That floodwater "covered the mountains to a depth of more than twenty feet (Genesis 7:20)" could as well be understood – in the context of other applications of the same words elsewhere in the Old Testament – to mean that the mountains were "drenched" and that water rose to a depth of twenty feet *against* the mountains.

Death and the Fall

The issue of "no animal death before the fall" is most pertinent to the Grand Canyon because of the flood geology claim that all fossils buried in the Earth's strata could only have perished after the fall, which introduced death to all creatures. Thus, fossils could not be the remains from hundreds of millions of years of death and burial. But nowhere does Scripture say that animal death resulted from man's sin. Genesis 3 does not make this claim, nor is it supported by commentary in the New Testament. Romans 5:12-13 tells us *"Therefore, just as sin entered the world through one man, and death through sin, and in this way death came to all men, because all sinned – for before the law was given, sin was in the world."* Here, the Apostle Paul is specific that death from sin applies to humans; he does not consider the death of animals to be consequential or relevant to his doctrinal point.

Some worry that the description of the physical world in the Bible is so different from what we know from modern science. Does the Bible contain scientific errors? The answer is No – if it was not *meant* to convey scientific information. Bible scholars and theologians reason that the scriptures were not written "to us," but "for us." Understanding the ancient Near East mindset helps provide context, because the people "to" whom the Bible was written did not care about the structure of the physical world in the way that modern scientists think about it; they cared about the *function* of the physical world. While the Genesis account of God making the world, and other biblical references to the natural world, do not use scientific language as we understand it, the Bible does explicitly claim that all that we see was created and is sustained by God's hand.

The Garden of Eden

The Bible does not claim, as flood geologists do, that all (or almost all) of the sedimentary rock on Earth formed in the Genesis deluge. In fact, this interpretation contradicts the literal biblical location for the Garden of Eden. The rich description of Eden in Genesis 2 is in near perfect concordance with the geography of Mesopotamia today, where four rivers merge at or near the Persian Gulf (Fig 2-5). Two of the rivers, the Euphrates and Tigris, are easily recognized by their ancient names. The Pishon is identified with an abandoned channel (the Wadi Batin) that flowed in antiquity (and that sometimes still flows today). The Gihon is probably the modern river Karun which still winds around and through the sedimentary rock of the Zagros Mountains of Iran. Natural resources and places known from the Arabian Gulf region, such as bdellium (a fragrant gum resin), bitumen (pitch), onyx, and gold, and

---

INFALLIBLE VIEW?

Young Earth Creationists insist that they are on secure footing with their view of nature because it is based on the revealed, infallible word of God. What few acknowledge, however, is that this view is not actually based on the assumption of the infallibility of the *Bible,* but on the assumption that their *interpretation and understanding* of the Bible is infallible.

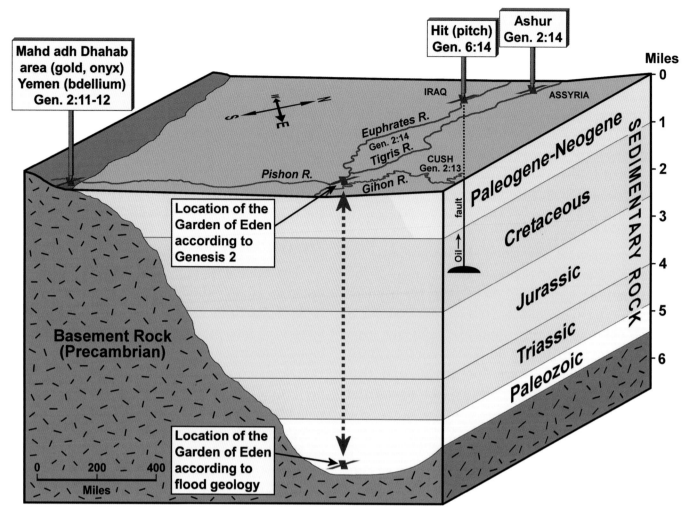

Figure 2-5. Schematic diagram of the Iraq/Arabia region and biblical location for the Garden of Eden (vertically exaggerated). Modified from C. A. Hill, The Garden of Eden: A Modern Landscape: *Perspectives on Science and Christian Faith*, v. 52, no. 1, pp. 31–46.

the still-present archeological site of Ashur – are also described in Genesis 2:10-14 with respect to the four rivers of Eden. Bitumen (pitch) was mined at Hit in ancient times, and it still is today. All of these clues further identify the Garden of Eden as being located on the Earth's surface *as we presently find it,* not on a catastrophically altered and buried landscape.

Recognition of the Garden of Eden as existing on a modern landscape presents a major (and ironic) conflict between what is a truly literal understanding of the text and what flood geologists believe about Earth history. The reason? There are six miles of sedimentary rock beneath the Garden of Eden/Persian Gulf. This is well

known by geologists because the area has been extensively drilled for oil, down to the crystalline basement (Precambrian rock). The pertinent question then becomes: How could Eden, which existed in *pre-flood* times, be located *on top of* over six miles of sedimentary rock supposedly deposited *during* Noah's flood? According to the flood geology model, Eden was flooded, buried under six miles of sediment, the crust was heaved violently upward and back downward, and somehow mineral and oil deposits appeared and rivers re-formed to *mimic* the old landscape, whereupon re-settlers gave rivers and places the same names – all to accommodate a scenario that the Bible never claims.

Flood Geology and the Grand Canyon

Flood geologists make the Grand Canyon the showcase for their view that most of the sedimentary rock on Earth formed during Noah's flood, only a few thousand years ago. Furthermore, they claim that almost the entire canyon was carved suddenly when one or more dammed lakes formed after the flood were breached, followed by the Colorado River carving the rest of the canyon over the last few millennia.

Different interpretations from flood geology and modern geology are compared in the two columns of the table below (Fig 2-6). Both sides claim these statements to be scientifically defensible, but the stark contrast in interpretations of the Earth's history reveal very different approaches to science. The conventional scientific approach puts all ideas and theories about the workings of nature to the test (not tests of whether nature is superintended by a divine Creator, but tests of whether nature has behaved in one particular way

or another). This approach starts with the questions and works forward to find answers. All the conclusions on the right-hand side of the table grew out of observations and testable hypotheses derived from studying the Earth and its surrounding solar system, with questions unfettered by preconceived notions of what the answers *should* be.

In contrast, flood geology starts with the answers and works backward to what questions should be asked. The conclusions of flood geology, on the left side of the table, start with a commitment to a set of particular, selectively literal interpretations of Scripture as "the answer," with ensuing questions designed only to support the predetermined conclusions. To flood geologists, these conclusions do not actually need to be testable because they are thought to represent revealed truth. Yet, flood geologists argue that their *geologic* interpretations *are* in fact testable by scientific investigation, so throughout this book we will evaluate the claims of flood geology on their scientific merits.

Flood Geology	Modern Geology
♦ Earth is about 6,000 years old.	♦ Earth is about 4.5 billion years old.
♦ Radiometric methods for the dating of geologic materials are flawed.	♦ Radiometric dating methods yield reliable dates on geologic materials.
♦ Noah's flood occurred about 4,500 years ago and was global over planet Earth.	♦ There is no surviving record of a global flood.
♦ It never rained on Earth before Noah's flood (some flood geologists no longer maintain this).	♦ Abundant evidence indicates rain throughout Earth's geologic history.
♦ Fossils in sedimentary rocks represent "all flesh" of Genesis 7:21, and the only terrestrial animal species to escape death were those saved on Noah's Ark.	♦ Fossils in sedimentary rocks are the remains of plants and animals that died and were preserved as the sediments turned into rock over millions of years.
♦ Most fossil-bearing sedimentary rocks on Earth, including most Grand Canyon strata, formed during Noah's flood in only one year's time.	♦ Sedimentary rocks have formed over hundreds of millions of years through the processes of sedimentation, compaction, and cementation.
♦ The Grand Canyon and Colorado River were formed rapidly when large post-flood lakes emptied catastrophically.	♦ The complex history of the Colorado River and the Grand Canyon is still being investigated, but the canyon's erosion involved millions of years.
♦ All pre-flood land features, including the four rivers of Eden, were covered by deposits from Noah's flood (i.e., sedimentary rocks).	♦ The Garden of Eden is described in Genesis as a modern landscape overlying sedimentary rocks.

Figure 2-6. Basic precepts of flood geology compared with those of modern geology.

Rainbow over the Inner Gorge near Granite Rapid. *Photo by Bronze Black.*

CHAPTER 3

TIME FRAME OF FLOOD GEOLOGY

by **Tim Helble** and **Carol Hill**

F LOOD GEOLOGISTS HAVE CONVINCED many people that the Grand Canyon's sedimentary rock layers are the products of a year-long global flood that occurred 4,000 to 5,000 years ago. To understand the flood geology model and its time frame, we need to start with the sequence of events described in the flood text of Genesis 7 and 8 (Fig 3-1).

Leading flood geologists divide all history into five periods, based on the description of events in Genesis: early creation week, pre-flood (day 3 of creation until beginning of the flood), early flood (first 150 days), late flood (day 151 to end of the

Figure 3-1. Time frames for the flood geology model.

Day 1	Genesis 7:11	On the 17th day of the second month, fountains of the great deep burst open and the flood gates of the sky were opened.
Day 1 to Day 40	Genesis 7:4; 7:12; 7:17	It rained for 40 days and 40 nights and the water increased and lifted up the Ark so that it rose above the Earth.
Day 40 to Day 150	Genesis 7:24	The Earth was flooded for 150 days or 5 months.
Day 150	Genesis 8:2	On the 17th day of the seventh month, the Ark rested upon the mountains of Ararat.
Day 224	Genesis 8:5	The waters decreased continually until the first day of the tenth month.
Day 284	Genesis 8:12	A dove was sent out a third time and did not return; waters abated enough for the dove to live on dry land.
Day 314	Genesis 8:13	The ground was drying, but not completely dry. Noah stayed in the Ark until the ground was dry.
Day 365	Genesis 8:14	The ground was completely dry. Noah left the Ark. If the first and last days are included, the total time of the flood event was 365 days, or the length of one solar year.

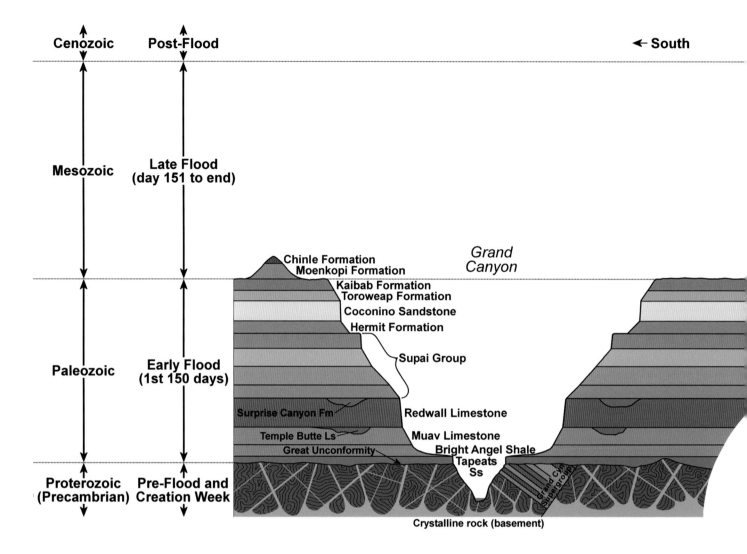

← South

Mesozoic   Late Flood
(day 151 to end)

Grand
Canyon

Chinle Formation
Moenkopi Formation

Kaibab Formation
Toroweap Formation
Coconino Sandstone
Hermit Formation

Supai Group

Paleozoic   Early Flood
(1st 150 days)

Surprise Canyon Fm   Redwall Limestone

Temple Butte Ls   Muav Limestone
Great Unconformity   Bright Angel Shale
Tapeats
Ss

Grand C.n
(Supergroup)

Proterozoic   Pre-Flood and
(Precambrian)   Creation Week

Crystalline rock (basement)

flood), and post-flood (Fig 3-2). According to this construct, the vast majority of all the Earth's sedimentary deposits – including those now comprising high mountains and those found tens of thousands of feet below the surface – would have been deposited, sorted, shifted, folded, tilted, and hardened into rock during or shortly after the flood.

Before we dive into the details of the time frame for flood geology, readers need to be aware of what is meant by the use of geologic "time" terms, such as *Precambrian, Jurassic,* or *Cretaceous.* The origin of these terms is covered in a later chapter, but for now it is enough to recognize them as names geologists ascribe to different time periods in the Earth's history.

Because these geologic time terms are routinely used by conventional geologists, one might naturally think they are rejected by Young Earth advocates. However, most flood geologists acknowledge the same basic *order* of the Earth's many layers and fea-

tures, so they generally accept the geologic naming system. The immense disagreement comes in the ages assigned to those time intervals and in the processes at work to create the rocks. The 4.5 billion years of conventional geology gets compressed into less than 10,000 years for the flood geologist.

So how does the Grand Canyon's geologic history fit within a year-long global-flood scenario? Most flood geologists acknowledge that the rocks of the Grand Canyon contain only part of the geologic history of the region. There used to be more rock layers above where the Grand Canyon is today, but those layers were eroded down to the one that we now see at the canyon rims – the top of the Kaibab Formation. The higher rock layers are still present north of the canyon, in what is known as the *Grand Staircase* (Fig 3-3, also see Fig 3-7 on page 37). When driving north into Utah, you "step up" through those higher Mesozoic rock layers like you are climbing

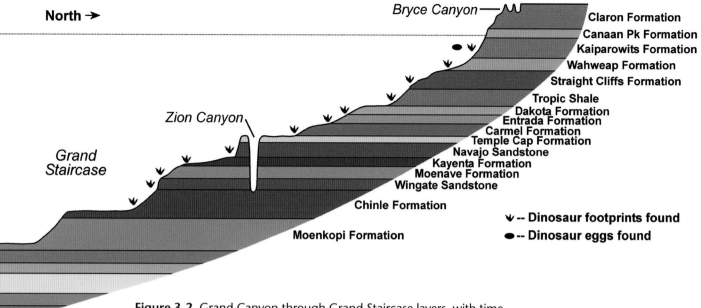

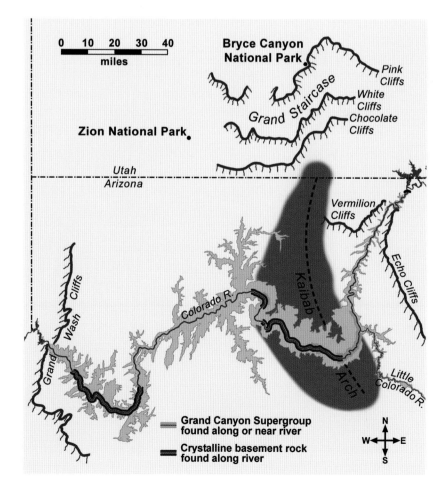

North →

Bryce Canyon ———

Claron Formation
Canaan Pk Formation
Kaiparowits Formation
Wahweap Formation
Straight Cliffs Formation
Tropic Shale
Dakota Formation
Entrada Formation
Carmel Formation
Temple Cap Formation
Navajo Sandstone
Kayenta Formation
Moenave Formation
Wingate Sandstone

Zion Canyon

Grand
Staircase

Chinle Formation

Moenkopi Formation

ꝟ -- **Dinosaur footprints found**
● -- **Dinosaur eggs found**

Figure 3-2. Grand Canyon through Grand Staircase layers, with time classification according to conventional geology and flood geology.

a staircase. Any view of the origin of rocks in this northern Arizona – southern Utah region thus has to account not just for the Grand Canyon layers being deposited from Noah's flood, but also for the thick sequence of layers in the Grand Staircase series of rocks.

Flood geologists don't all agree on which layers should be assigned to each flood-related period. In fact, they actually disagree to the point that if all flood geologists who have ever published on the subject had equal authority in the Young Earth community, one would be hard-pressed to decide which, if any, of the layers in the Grand Canyon/Grand Staircase series were deposited by Noah's flood. Nevertheless, the flood geologists with the greatest influence have classified the Grand Canyon and Grand Staircase layers according to the time frames written on the left side of Figure 3-2.

Figure 3-3. Map of Grand Canyon–Grand Staircase area. Key locations and features mentioned in the text are labeled.

On the following pages, the rock layers and features of the Grand Canyon and Grand Staircase are divided into sequential time periods, numbered one

0 10 20 30 40
miles

Bryce Canyon National Park

Pink Cliffs
White Cliffs
Chocolate Cliffs

Grand Staircase

Zion National Park.

Utah
Arizona

Vermilion Cliffs

Echo Cliffs

Grand Wash Cliffs

Colorado R.

Kaibab Arch

Little Colorado R.

▬ **Grand Canyon Supergroup** found along or near river
▬ **Crystalline basement rock** found along river

N
W ← → E
S

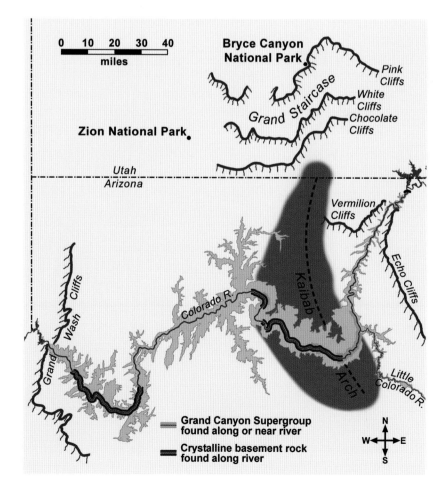

through five, with descriptions of their purported history according to flood geologists. In the next chapter, the same rock groupings and numbering sequence are used to directly contrast the conventional geology and flood geology interpretations.

1. Crystalline Basement Rock:
Early Creation Week

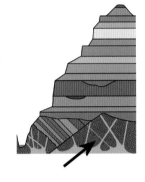

Underlying all sedimentary rock in the Grand Canyon is a "basement" of metamorphic rock with many igneous intrusions (Fig 3-4). This crystalline rock (rock with visible, interlocking crystals) first appears in the Upper Granite Gorge of the eastern Grand Canyon, beneath Grandview Point, and continues to the west until it disappears beneath sedimentary layers in the central Grand Canyon. It reappears in the far western Grand Canyon (see markings on Fig 3-3 map, page 33).

Igneous rock forms when magma (melted rock) cools and crystallizes. Metamorphic rock forms when heat and/or pressure changes the texture and mineral composition of sedimentary or igneous rock, while the rock itself remains solid.

WHAT PROCESS FORMED THE SUPERGROUP ROCK SO RAPIDLY?

The Supergroup layers are almost as thick as the overlying layers of the Grand Canyon and Grand Staircase combined, and they have a similar assortment of sedimentary rock types. Given the 1650-year window allowed by flood geologists for deposition and hardening of the Supergroup, its layers had to form at a phenomenal rate averaging 7 ft/yr, during a time when some flood geologists say it didn't even rain. If a massive flood was required to deposit the layers of the Grand Canyon and Grand Staircase, how were thousands of feet of Supergroup layers deposited *without* a flood?

Figure 3-4. Upper Granite Gorge. *Photo by Tom Bean.*

We know sedimentary rock can change into metamorphic rock on a large scale because there are places in the world where we can find unaltered sedimentary rock transitioning laterally into more and more altered metamorphic rock.

Flood geologists don't provide many specifics on the origin of large areas of metamorphic rock such as those seen in the Grand Canyon. They do acknowledge that these started as unaltered sedimentary and igneous rock, but argue that the subsequent metamorphism was extremely rapid, caused either by shifting crust during the creation week when land was being formed, or during Noah's flood when parts of the Earth's crust were violently set into motion. For the Grand Canyon, flood geologists appear to favor the first explanation.

Figure 3-5. Tilted Grand Canyon Supergroup rocks. Elevation range in the photo is about 2100 feet.
Photo by E.D. McKee, Courtesy U.S. Geological Survey.

Nankoweap Fm.
Cardenas Lava
Dox Formation

2. Grand Canyon Supergroup: *Pre-Flood*

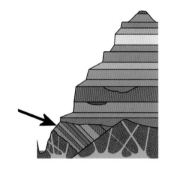

A stack of tilted sedimentary rocks and lava, known as the Grand Canyon Supergroup, overlaps on top of the crystalline basement rocks. If you could "untilt" the entire Grand Canyon Supergroup, its total thickness would be about 12,000 feet. Looking north from Lipan Point, Moran Point, or Desert View, on the South Rim, you can get an idea of the thickness of just three of the Supergroup's layers: Dox Formation, Cardenas Lava, and Nankoweap Formation (see Fig 3-5).

The most influential flood geologists classify the Grand Canyon Supergroup as pre-flood material that was deposited and hardened into rock some-

time between the third day of creation and Noah's flood some 1,650 years later. They generally do not address the mechanism for amassing and hardening more than 12,000 feet of sedimentary rock *before* the flood. The Supergroup layers contain mostly fossils of stromatolites (colonies of single-cell organisms). Flood geologists believe that organisms similar to modern ones lived and died during this time, so the absence of any ferns, flowering plants, fish, birds, reptiles, mammals, amphibians, sharks, or other multicellular organisms in over 12,000 feet of Supergroup sedimentary rock also remains unexplained.

3. Tilting and Erosion of the Supergroup, and Deposition of Overlying Horizontal Layers: *Early Flood*

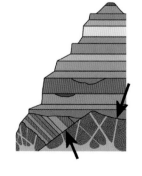

Flood geologists identify the early flood period as starting when "the fountains of the great deep burst forth, and the windows of the heavens were opened" (Genesis 7:11), and ending at Day 150 (Genesis 7:24; 8:3). According to the flood model, a single "supercontinent" existed before the flood that was violently fractured and thrust apart, eventually leading to the Earth's current arrangement of continents.

A video at the Creation Museum in Kentucky shows huge tsunamis being generated by this break-up – waves large enough to be visible from space. In this view, giant tsunamis circled the planet in a matter of hours and wiped out almost everything in their paths (Fig 3-6). Rapid break-up of the supercontinental plate exposed cool seawater to hot magma below, causing plumes of steam to jet high in the sky. These plumes were "the fountains of the great deep" and the source of water for "the floodgates of the sky."

An alternative explanation provided by a few flood geologists calls for a solid, unbroken crust of rock covering the pre-flood Earth, which in turn sat on top of a layer of water. In this version, cracks, formed in the crust at the start of the flood, allowed super-pressurized water from below to jet upwards at supersonic speeds (fountains of the deep). Massive salt and limestone beds were deposited where this mineral-rich water flowed into the sea.

The entire Grand Canyon Supergroup is said to have been tilted and faulted by tectonic upheavals during the early days of the flood, followed by violent erosion of the tilted rock layers to create a nearly flat surface. In some places, that erosion would require over 12,000 feet of pre-flood sedimentary rock to be stripped away to expose the underlying crystalline rock. The first modern flood geologists assumed this violent erosion at the start of the flood would provide enough sand, silt, clay, and other material needed to form the Earth's sedimentary rock layers, totaling over 100 million cubic miles in volume.

However, flood geologists now recognize that erosion at the start of the flood wouldn't be able to supply enough sediment, and instead suggest that significant stockpiles of loose sediment already existed in various areas of the pre-flood Earth. This view requires that some stockpiles were protected from early flood waters, so they would still be available for transport later in the flood.

Though the initial violence of the flood is claimed to have ripped apart rocks and caused massive erosion, flood geologists say that only days later these same violent floodwaters started laying down thick deposits of sediments washed in from distant regions. The mile-thick accumulation of roughly horizontal layers in the Grand Canyon, from the Tapeats Sandstone all the way up to the Kaibab Formation, is claimed to be from early flood deposits. Flood geologists have proposed mechanisms for how they believe many of these distinct layers could have formed, but detail is often lacking with regard to how the unique composition, structures, and fossils of each layer could have come about in a matter of days.

4. The Grand Staircase (Grand Canyon to Bryce Canyon): *Late Flood*

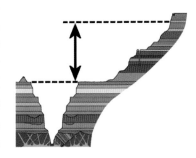

The late flood period starts on Day 151 and ends when the one-year flood is over. Flood geologists claim this second part of the flood was responsible

Figure 3-6. Illustration of the "onset of the flood."
Watercolor by Jeff Dombek.

Figure 3-7. The Grand Staircase looking north from the air, with clear views of the Vermilion Cliffs, White Cliffs, and distant Pink Cliffs. *Photo by Scott Braden.*

for depositing the layers of rock seen in the Grand Staircase north of the canyon (Fig 3-7, see also map and geologic diagram on pages 32-33). During this period the fountains of the great deep were shut off, and rapidly moving continental plates began to slow down. However, high mountains and deep trenches continued to form violently, earthquakes rumbled, and swift currents raged.

Leading flood geologists understand the wording found in Genesis 8:3 and 8:5, *"the water receded steadily from the earth…"* and *"the water decreased steadily…,"* to mean that the late flood period was characterized by waters rushing back and forth with an action resembling violent tidal fluctuations. The resulting currents were forceful enough to strip away thousands of feet of sediment in some areas and create equally huge deposits in other places.

According to the flood geology model, the thick sequence of Grand Staircase rocks (over 5,000 additional feet) originally covered the Grand Canyon area as well. As the floodwaters receded, swift currents stripped away thousands of feet of this sediment in the Grand Canyon area, removing layers all the way down to the Kaibab

Formation, where the canyon rims are today. To the north of the Grand Canyon area, the late flood layers were not cleared away, and eventually even more layers were added on top. In the midst of all that violence, flood geologists believe there were numerous breaks in activity, with many life forms surviving months into the deluge. Many kinds of dinosaurs, for example, somehow survived the initial onslaught of continent-sweeping tsunamis and months of inundation that followed to leave footprints in hundreds of different freshly deposited layers, as well as many egg-filled nests (Figs 3-8, 3-9).

5. Bryce Canyon and Higher Rock Layers: *Post-Flood*

Flood geologists maintain that heightened geologic activity continued after the flood. Until that activity decreased, local or regional sediment

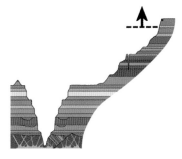

deposition continued for many years after the flood. Opinions in the flood geology community vary widely on exactly where the late-flood/post-flood boundary should be defined, with no clear evidence supporting any particular position. Since there is no consensus, flood geologists seem to have agreed to defer their final decision on this matter to the future.

One published source, *Grand Canyon, Monument to Catastrophe*, identifies the Cenozoic Claron Formation as the first post-flood layer in the Grand Canyon region (see map and geologic diagram on pages 32-33). The Claron Formation is a lake deposit, supposedly formed in one of the many lakes that existed in the area after the flood. It contains fossils of freshwater fish and snails. The unusual pillars, called *hoodoos*, and steep gullies of Bryce Canyon are carved into this formation (Fig 3-10). There are thousands more feet of "post-flood" sediments above the Claron Formation in the western United States. Great sequences of "post-flood" rock are also found in other parts of the world, all deposited without the benefit

THE DINOSAUR ESCAPE HYPOTHESIS

Certain flood geologists claim that some dinosaurs could have survived the 150-day early flood period and established nesting grounds before being wiped out during the late flood period. Consider the implausible sequence of events required for this "escape hypothesis." Early in the flood, giant waves circled the Earth, scouring parts of continents and in other places dumping sediment in massive deposits thousands of feet thick. Changing currents and continental upheaval caused these newly formed deposits to rise above sea level, while new waves raged over previously unflooded land to sweep off dinosaurs. Some dinosaurs swam or clung to floating debris long enough to gain footing on the freshly uplifted muck at multiple places around the globe, with sufficient numbers still alive at each location to establish nesting grounds. But shifting currents and tectonics then redirected waves to bury these hapless communities repeatedly within several more thousand feet of sediment (Fig 3-2).

Figure 3-8. Tracks of a theropod dinosaur at Flag Point on the top of the Vermilion Cliffs, Utah. *Photo by Scott Thybony.*

Figure 3-9. *Troodon* egg clutch from Teton County, Montana. *Photo by Tim Evanson. Photo courtesy of Museum of the Rockies. Note: Even though this came from Montana, it was found at the same level in the stratigraphic column as higher parts of the Grand Staircase.*

of catastrophic upheavals and global inundation. Finally, the carving of the Grand Canyon (which we will cover in later chapters) is also said to have occurred during the post-flood period.

Does a Flood Geology Time Frame Sound Reasonable?

Gigantic tsunamis circling the Earth in a matter of hours, land rising violently above sea level and then being thrust back down below, huge pre-flood stockpiles of sediment waiting to be transported by the flood, astronomical rates of sediment transport and deposition, and dinosaurs surviving and nesting during months of violent, global inundation — scenarios that are all based on a supposedly "literal" reading of Genesis. Is any of this plausible? In the next chapter, we'll compare this flood geology view with the conventional geology view.

THE PROBLEM OF FRESHWATER FISH

Most freshwater fish quickly die when suddenly placed in seawater. In a global flood sending ocean water over the Earth and lasting a year, all but the most salt-tolerant fish should have died. Why then do we have fossil evidence of abundant freshwater fish in many "post-flood" lake deposits?

Figure 3-10. Bryce Canyon, carved in the lake limestone of the Claron Formation, the highest rock unit of the Grand Staircase. *Photo by Tim Helble.*

Vishnu Temple in the eastern Grand Canyon, looking south from Cape Royal on the North Rim.
Photo by Mike Buchheit.

CHAPTER 4

TIME FRAME OF MODERN GEOLOGY

by **Carol Hill** and **Stephen Moshier**

THE MODERN GEOLOGIC UNDERSTANDING OF THE history of the Grand Canyon, and of the Earth, is vastly different from what is proposed by flood geologists. The divergence is much more than length of time. Great differences also exist in the processes and environments understood to form each layer or feature, and in how simple or fantastic the explanations must be to account for the data. In this chapter, we will briefly summarize the history told by the rocks of the Grand Canyon and Grand Staircase from the conventional geologic perspective. At this stage, we won't go into much detail about how we know that history; the explanations will come in subsequent chapters.

We'll start with the big picture – the age of planet Earth – so we can better understand how the Grand Canyon layers fit into the larger story.

Age of Planet Earth

Modern geologists and astronomers place the overall age of our planet Earth at about 4.5 billion years, based on multiple radiometric dating methods applied to the oldest known rocks (Earth and Moon) and meteorites. These methods come with no evolutionary or humanistic assumptions – just basic physics applied to atoms and rocks.

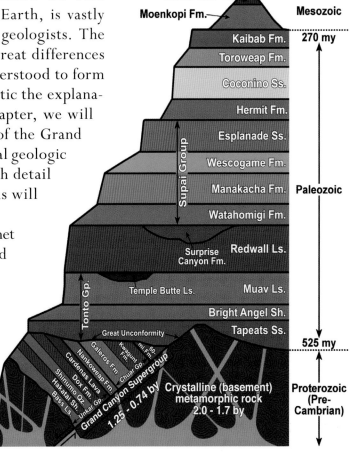

Figure 4-1. Grand Canyon geologic column. Ages for the Paleozoic layers represent the ages of the Tapeats and Kaibab Formations, not the beginning and end of the Paleozoic Era. (by=billion years; my=million years)

FORMATION OF COLORADO PLATEAU ROCK LAYERS AND THE CARVING OF THE GRAND CANYON

(Figure 4-2)

1.75 billion years ago

1.70 billion years ago

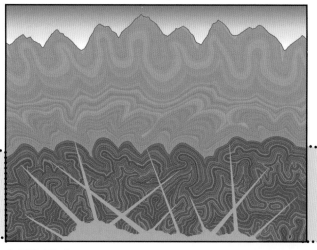

SCENE 1
Accumulation of thick sequence of sediments and volcanics where tectonic plates collide, (where one plate slips beneath the other, it is known as *subduction*).

SCENE 2
Converging plates uplift mountains, deformation and metamorphism of deeply buried sediments and volcanic rock, and intrusion of granite plutons, dikes, and sills.

525 million years ago

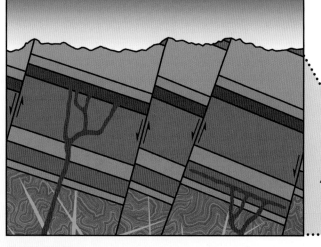

525 to 270 million years ago

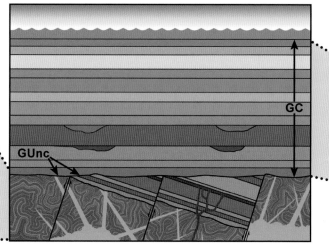

SCENE 5
Precambrian rocks tilted and faulted into separate blocks.

SCENE 6
Erosion of Precambrian rocks, and deposition of more than 4,000 feet of Paleozoic strata. GC = Grand Canyon layers, GUnc = Great Unconformity.

The drawings on these two pages depict eight simplified snapshots or "scenes" of the formation and erosion of the rock layers seen today in the Grand Canyon. They are simplified representations of the Earth's crust in the general region of what is today the Colorado Plateau. (Layers and colors match the labeled figure in Chapter 3, pages 32-33.)

1.25 billion years ago

1.25 billion to 740 million years ago

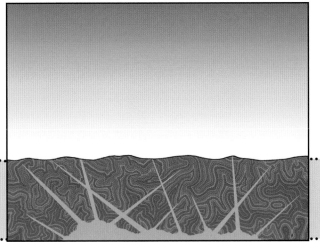
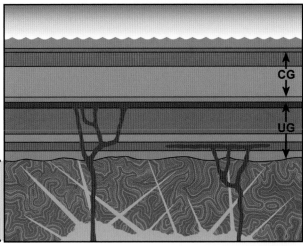

SCENE 3
Erosion of the mountains to a relatively flat surface, exposing metamorphic crystalline basement rock.

SCENE 4
12,000 feet of Supergroup rocks laid down on eroded surface. Intrusion of dikes and sills.
CG = Chuar Group, UG = Unkar Group.

270 to 65 million years ago

Today

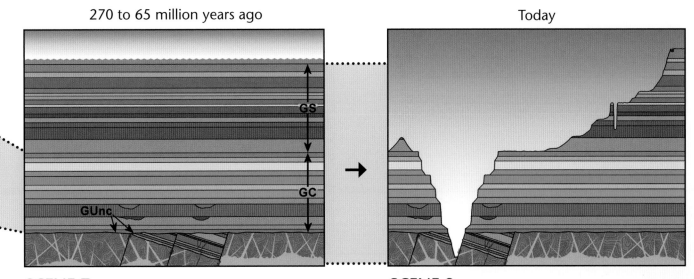

SCENE 7
Additional deposition of Mesozoic Grand Staircase strata (5,000 to 10,000 feet). GS = Grand Staircase layers, GC = Grand Canyon layers, GUnc = Great Unconformity.

SCENE 8
Removal of Mesozoic strata over Grand Canyon area and carving of Grand Canyon.

Figure 4-3. View of Grand Canyon from the Tonto Trail, with formations labeled. Hermit's Rest is in the upper right corner. *Photo by Tim Helble.*

In any given spot on Earth, the ensuing years after the planet formed witnessed innumerable environmental changes, some fast and some slow. Because those changes included erosion as well as deposition, no place on our planet has a continuous, uninterrupted record of rocks from the beginning to the present (although there are several places where sediments from every geologic period are found in one thick stack). The Grand Canyon has more of this record spectacularly exposed than perhaps anywhere else – which is why we and the flood geologists find ourselves so interested in this locale (Figs 4-1, 4-3).

So what story do geologists see preserved in the rocks of the Grand Canyon? The division of rock layers into the flood geology time frame does not translate into the most logical division of layers for the conventional geologic model, but to make the comparison with the flood geology model easier to follow, we will use the same groupings of rocks and features with the same numbering identified in Chapter 3, *Time Frame of Flood Geology*. The eight-panel sequence on the previous two pages illustrates scenes from major episodes in the history of the Grand Canyon (Fig 4-2). The remainder of this chapter provides more detailed explanations on how the succession of rocks shown in

these eight scenes led to what we see today in the Grand Canyon.

1. Crystalline Basement Rock: *Proterozoic*

The oldest rocks in the Grand Canyon are exposed in the Inner Gorge. They consist of metamorphic rock, so named because the rock has

SCENE 1: Accumulation of thick sequence of sediments and volcanics where tectonic plates collide, (where one plate slips beneath the other, it is known as *subduction*). (Middle Proterozoic)

Figure 4-4. Granite plutons, dikes, and sills intruded into the Vishnu Schist in the Inner Gorge. *Photo by Gary Ladd.*

Figure 4-5. Pink-colored dikes of Zoroaster Granite cutting across the darker-colored Vishnu Schist. *Photo by Wayne Ranney.*

undergone a metamorphosis or transformation from one type of rock into another type of rock. Burial deep in the Earth's crust created intense pressures and temperatures that caused the rock now seen in the Inner Gorge to change into new types of rock. The main kind of metamorphic rock exposed in the Inner Gorge is schist, which appears to have been sedimentary rock and volcanic lavas before alteration.

Several massive bodies of granite, called plutons, are also found within this complex of metamorphic rocks. Granite pegmatites (igneous rocks with very large crystals) form *dikes* and *sills* that were injected from the plutons below into the older metamorphic rocks above (Figs 4-4, 4-5). (Intrusions that cut across existing layers are called dikes; intrusions that squeeze between and run parallel to existing layers are called sills.)

Radiometric dating (see Chapter 9) indicates that the granites intruded about 1.7 billion years ago. The older surrounding rock had to be deposited and buried before being metamorphosed and intruded by the granite. Some minerals in the schist come from rocks formed up to 3.3 billion years ago – reaching nearly three quarters of the way back through Earth's 4.5 billion year history!

Geologists have determined that the oldest rocks in the Grand Canyon got their start in an offshore volcanic-island setting where one tectonic plate subducted under another (Scene 1), resulting in a string of volcanic islands – much like those off the southwest coast of Alaska today. Thick deposits of sand, mud, and lava flows (inside the outlined box in Scene 1) formed off the flanks of the volcanic islands.

Forty million years later, the volcanic islands collided against what was then the continental edge of North America. The collision created a

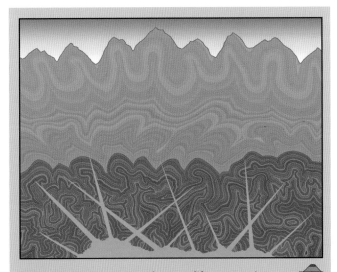

SCENE 2: Converging plates uplift mountains, deformation and metamorphism of deeply buried sediments and volcanic rock, and intrusion of granite plutons, dikes, and sills.

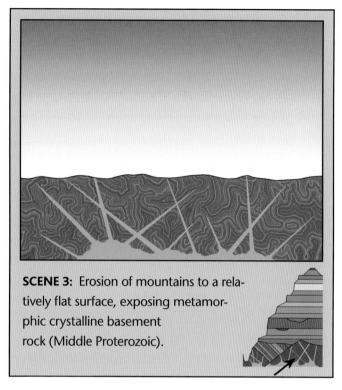

SCENE 3: Erosion of mountains to a relatively flat surface, exposing metamorphic crystalline basement rock (Middle Proterozoic).

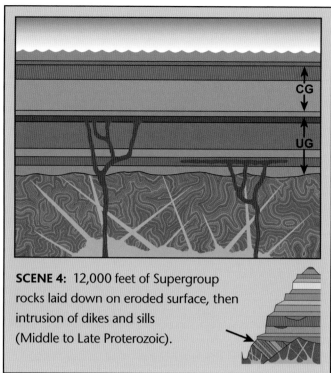

SCENE 4: 12,000 feet of Supergroup rocks laid down on eroded surface, then intrusion of dikes and sills (Middle to Late Proterozoic).

tall mountain chain with the older volcanic and sedimentary rock created in Scene 1 buried miles deep under the mountains. The signs left in these rocks, such as the size, shape, and composition of crystals, are consistent with a history of eventual burial deep enough to begin alteration of the minerals' composition and character, and ultimately to turn them into metamorphic rock. Then, deep beneath the surface, magmas from hotter rocks below intruded the metamorphic rocks to form granite plutons, dikes, and sills about 1.7 to 1.66 billion years ago (illustrated in Scene 2 as pink rock with spikes). These are the Vishnu Schist and granite dikes exposed in the Grand Canyon's Inner Gorge.

Erosion of the mountains over some 400 million years exposed the once deeply buried metamorphic and igneous rocks at the Earth's surface. Erosion continued until the surface was close to sea level (Scene 3).

Figure 4-8. Stromatolite fossil from the Chuar Group. *Photo by Doug Powell.*

above: **Figure 4-6.** Modern stromatolites in Shark Bay, Australia. *Photo by Paul Copper.* *right:* **Figure 4-7.** Eroded cross section of a stromatolite, showing what the inside layers look like. *Photo by Gregg Davidson.*

2. Grand Canyon Supergroup:
Proterozoic

Stretching of the Earth's crust created a new basin for sediment to accumulate over the exposed surface of the metamorphic-igneous bedrock. This resulted in the deposition of the Grand Canyon Supergroup (Unkar and Chuar Groups – the colored, horizontal layers) from about 1.25 billion to 740 million years ago (Scene 4).

Many of the Unkar rocks display ripple marks and mud cracks, indicating that their original deposition was in shallow water that intermittently dried up. The alternating layers of limestone and sediments in the Unkar Group are consistent with periods of rising and falling sea level (or subsidence and uplift of the land). The fossils of an early life form known as stromatolites are found in the Unkar Group. Stromatolites are a type of cyanobacteria that form disk-shaped colonies, which often grow upward to create short pedestals. (Smaller, free-rolling versions are technically known as *oncolites*. For simplicity, all are referred to as stromatolites in this book.) Stromatolites are some of the earliest life forms in the fossil record anywhere on planet Earth. Stromatolites are found in strata right up to the present, including in the coastal waters of Western Australia and the Bahamas (Fig 4-6). While the Unkar sedimentary rocks cannot be directly

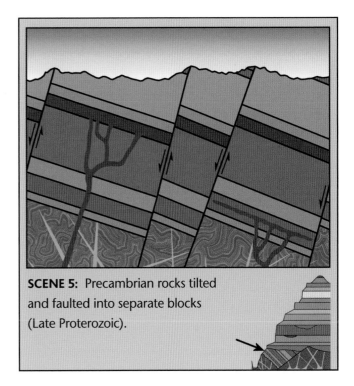

SCENE 5: Precambrian rocks tilted and faulted into separate blocks (Late Proterozoic).

dated, the group is underlain by a volcanic ash bed, which yields a radiometric age of 1.25 billion years, and capped by the Cardenas Lavas (the reddish dikes and lava flows), with a radiometric age of about 1.10 billion years. The Cardenas Lava intruded and flowed over the Unkar Group about 1 billion years ago before the Chuar Group was deposited.

The Chuar Group of sedimentary strata overlies the Unkar Group and contains fossils of very primitive

Figure 4-9. Tilted Precambrian Shinumo Quartzite, Hakatai Shale, and Bass Formation under horizontal Paleozoic layers in the eastern Grand Canyon. *Photo by Gregg Davidson.*

Figure 4-10. Great Unconformity, in Blacktail Canyon, River Mile 121, where horizontally bedded Tapeats Sandstone overlays the darker Vishnu Schist. *Photo by Howard Lee.*

life forms, including stromatolites (Figs 4-7, 4-8). Rocks of similarly ancient age around the world also lack fossils of any greater diversity. The most reasonable explanation for this lack is that any other life forms besides the very simplest were not yet present on Earth.

3. Tilting and Erosion of the Supergroup, and Deposition of Overlying Layers: *Late Proterozoic through Paleozoic*

The flood model lumps tilting of the Supergroup, erosion to form the Great Unconformity, and deposition of all the overlying Grand Canyon layers into the first 150 days of the flood year. Conventional geologic understanding divides these into two very different time frames, separated by hundreds of millions of years.

Continued stretching created faults that broke the crust into tilted blocks, where some blocks of rock moved up and down relative to adjacent blocks (Scene 5). The Grand Canyon Supergroup rocks were cemented and hardened before they were

faulted. Uplift of the fault blocks resulted in another long episode of erosion lasting 200 million years.

By about 525 million years ago, erosion had left a low-lying surface close to sea level, but with some high ridges of more resistant rock (some up to 800 feet in relief). The bedrock exposed at the surface included remnants of the Supergroup fault blocks and stretches of older metamorphic and igneous rocks. Geologists call this vast erosion surface, or contact between older bedrock and overlying younger sedimentary rock, the *Great Unconformity.*

While flood geologists see no significant time gap represented by the Great Unconformity (erosion and deposition both caused by the same flood), modern geologists recognize an enormous gap in time. The magnitude of that gap depends on what part of the canyon you are in — whether the overlying rocks rest on the oldest schist and granite, or on the exposed edge of a younger layer in the Supergroup (Fig 4-9). Where the Tapeats Sandstone sits directly on the schist and granite, the gap is over a billion years (Fig 4-10)!

Over the next 255 million years, during the Paleozoic Era, the sea alternately advanced and receded across the region as the crust subsided, leading to the accumulation of thousands of feet of sedimentary layers (Scene 6, Fig 4-11). Streams

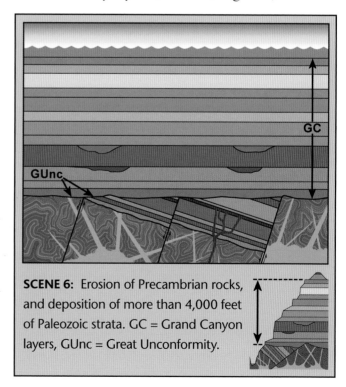

SCENE 6: Erosion of Precambrian rocks, and deposition of more than 4,000 feet of Paleozoic strata. GC = Grand Canyon layers, GUnc = Great Unconformity.

Figure 4-11. The Paleozoic layers of Grand Canyon near River Mile 122. *Photo by Tim Helble.*

cut channels into the rock (represented by the four bowl-shaped areas near the middle of the diagram) during times when the sea retreated. Deposition and erosion of rock occurred periodically until about 270 million years ago when the Kaibab Formation (now the top rock layer in the Grand Canyon) was deposited in a shallow sea.

The alternating layers of sandstone, shale, and limestone are simply explained by cycles of rising and falling sea level (often due to the crust subsiding or uplifting), with quartz sand collecting along beaches and in shallow seawater, in riverbeds, and in desert sand dunes; mud accumulating in protected bays or deeper, relatively still water; and layers of limey sand and mud forming in warm, clear water where organisms with shells grew and died.

Many of these layers bear evidence of erosion (more unconformities) following exposure above sea level, with abrupt transitions to different deposits or life forms when sea level rose again and deposition resumed. Some layers, such as the Redwall Limestone, were raised above sea level after formation long enough for an extensive cave network to develop, formed by

percolating rainwater and groundwater flow, and later filled in by overlying sediments.

A wealth of fossils is preserved in the canyon rocks. All, without exception, are recognized by paleontologists (geologists that study fossils) as being Paleozoic organisms (541 to 252 million years ago). Common organisms found in younger sediments worldwide, such as birds, dinosaurs, mammals, and flowering plants, are completely absent in Grand Canyon rocks.

4. The Grand Staircase (Grand Canyon to Bryce Canyon): *Mesozoic*

The rock layers of the Grand Staircase north of the Grand Canyon (Fig 4-12), once extended over the Grand Canyon area (Scene 7). These layers were also deposited from fluctuating seas, though during the later Mesozoic Era, beginning at about 252 million years ago and continuing until about 65 million years ago – the end of the reign of dinosaurs.

In the Mesozoic layers of the Grand Staircase, above the older Grand Canyon layers, we find fossils that are absent in the Grand Canyon: dinosaurs, marine reptiles, flying reptiles, flowering plants, and, eventually, early mammals in the upper layers. All are recognized around the globe as Mesozoic organisms (252 to 66 million years ago). The lack of any mixing between marine and land organisms

Figure 4-12. Grand Staircase. *Photo by Wayne Ranney.*

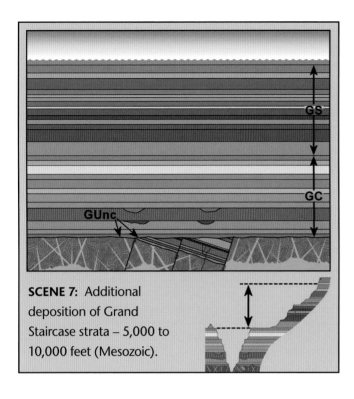

SCENE 7: Additional deposition of Grand Staircase strata – 5,000 to 10,000 feet (Mesozoic).

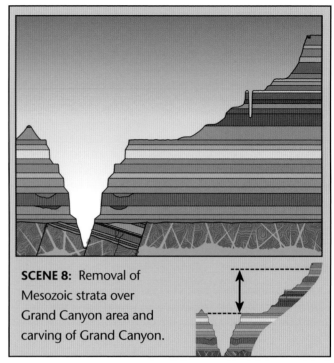

SCENE 8: Removal of Mesozoic strata over Grand Canyon area and carving of Grand Canyon.

in the different layers, and the presence of preserved dinosaur nests containing eggs, are evidence of normal (not megaflood) processes at work.

5. Bryce Canyon and Higher Rock Layers: *Cenozoic*

The highest unit in the Grand Staircase is the Claron Formation, a lake limestone that forms the fantastic hoodoos (rock columns) at Bryce Canyon National Park (Fig 4-13). Elsewhere in North America and around the world, there are many more rock layers that are even younger than the Claron. These layers likewise contain ample evidence of the same processes evident in lower layers, and they too contain unique fossils. Here we find bison, mammoths, flowering plants, and a complete absence of the many extinct life forms – such as trilobites and dinosaurs – found in lower layers of Paleozoic and Mesozoic rock.

The Grand Canyon was carved into these stacked rock layers when the Colorado Plateau up-lifted during the Cenozoic Era (Scene 8), causing the sea to retreat from the region. The age of the earliest carving of the Grand Canyon is much in debate. While part of it probably started in response to this uplift, the majority of canyon carving by the

Colorado River happened from about 6 million years ago to the present.

Flood geologists lump the deposition of the Claron Formation, in the Grand Staircase (Fig 4-13), and the carving of the Grand Canyon into the same brief post-flood time frame (the last 4,500 years). Modern geologists place the Lake Claron deposits in the time period when the rocks of the Grand Canyon and Grand Staircase were being uplifted to eventually form the Colorado Plateau. The timing of the uplift, and the carving of Grand Canyon itself, will be covered in later chapters.

Sorting Things Out

Young Earth advocates contend that we all are looking at the same evidence but with different interpretations resulting simply from differing worldviews: biblical versus naturalistic. Yet most geologists, including most Christian geologists, argue that flood geologists are not considering all the evidence or are selectively reporting only the evidence that supports their interpretations. How does one find his or her way from these disparate positions? Can anyone know what really happened in the unobserved past? We believe we can answer that – just keep reading!

Figure 4-13. Hoodoos of the Claron Formation in Bryce Canyon. *Photo by Bronze Black.*

Central Grand Canyon at sunset. *Photo by Bronze Black*

PART 2

How Geology Works

In Part 1, we explained what flood geology is and how it vastly differs from modern geology in its time frame and interpretations. In Part 2 we will explain the basic principles by which the science of modern geology "works" – that is, how geologists arrive at their conclusions. "How geology works" is divided into three subsections: (1) sedimentary rocks (how sedimentary rocks and structures form, and how the present is used to understand the past; Chapters 5–7), (2) time (how we know the order of events or the age of something; Chapters 8–10), and (3) plate tectonics (forces at work that lift mountains and deform rock; Chapters 11–12).

SEDIMENTARY ROCKS

Sedimentary rocks are emphasized in the next three chapters, in part because they are the rocks that make up the bulk of the Grand Canyon and the Grand Staircase to the north, and also because much of what we know about Earth's overall history is contained in sedimentary rocks. In previous chapters, we briefly described igneous rock (formed from magma) and metamorphic rock (altered by heat/pressure) in the canyon. The third type, sedimentary rock, forms by one of two primary processes: by deposition and cementation of grains transported from some other location, or by precipitation of minerals directly from water. The latter can include collections of shells formed by organisms living in the water.

In Chapters 5–7, we will explain how geologists can recognize in sedimentary rocks the depositional processes at work and the environmental settings that lead to the formation of rock layers. They do this by comparing the rocks with similar sedimentary deposits forming on the land surface and under the sea today. The amazing similarities between ancient rocks and modern sedimentary deposits are among the most compelling reasons why geologists reject the flood geology model of rapid and catastrophic formation of the rock strata in the Grand Canyon and Grand Staircase.

A Navajo Sandstone exposure, known as "the Second Wave," north of the Grand Canyon (part of the Grand Staircase). *Photo by Mike Koopsen.*

CHAPTER 5

SEDIMENTARY ROCK TYPES AND HOW THEY FORM

by **Stephen Moshier, Tim Helble,** and **Carol Hill**

W̲E FINISHED CHAPTER 4 WITH A question. Can anyone ever know what actually happened in the unobserved past? Put another way, when we stand in front of a cliff face with layers of different kinds of rock – some tilted, some folded, some faulted and offset, some with fossils and some without, and some partially eroded – how can we hope to figure out such a complex history? Answering this question for a particular location may take years of careful study, but the basic tools in a geologist's toolbox are actually fairly simple. We start by observing how sediments form today.

Sedimentary Rocks Are Made from Sediments

Our first observation is the perhaps self-evident reminder that sedimentary rocks derive from the

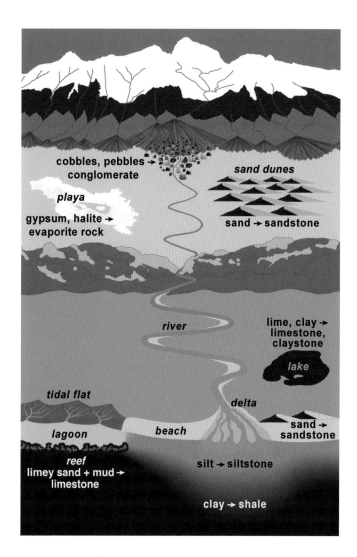

Figure 5-1. Idealized illustration depicting many common environments where different kinds of sediments are deposited. Each deposit has characteristics that are unique to that environment. For example, deposits of sand from desert dunes look very different from sand deposited along a seashore. The diagram shows what kinds of sedimentary rock are created in these different settings.

accumulation of sediments. What likely is less obvious is the fact that sediments can form in very different ways. It all depends on where the sediments came from, and where and how they were deposited. This is illustrated in the figure on the previous page, which shows some of the typical environments or settings where sediments are formed and deposited (Fig 5-1). We'll refer back to this illustration repeatedly throughout this chapter.

We can start in the mountains, where bedrock has been lifted high and exposed to the weathering and erosional forces of glaciers, running water, and landslides. Naturally, one would expect the sediment formed here to be very coarse and similar to the bedrock it comes from – composed of boulders, cobbles, and some sand. The newly made sediment is transported by landslides and mountain streams out onto the valley floor, where it resides temporarily in broad fans of rock debris. Rivers flushing through the fans carry the sediment away from the mountains. River transport of the sediment breaks it down into gravel and sand particles that move along the bottom of the stream. Finer silt and clay are carried suspended in the water, along with dissolved minerals. Large and small particles are transported where the stream flows fast. As the stream slows, the gravel drops out first, next the sand, then the suspended silt, and finally the clay, if flow virtually stops. On a small scale, this is observed on the inside bend of a stream, where slower-moving water commonly builds up sand bars. The distance that particles travel influences their shape and composition. Long pathways or fast water rounds and smooths larger fragments (pebbles and cobbles) by tumbling them along the bottom. A long trip down a stream also allows time for soft minerals to be worn away, or for other minerals to be dissolved or altered into fine clay minerals.

Rivers are like conveyor belts, moving sediment from the mountains to the sea. However, there may be stops along the way where sediment can accumulate in lakes or be sifted by desert winds. Sediments that are carried to the sea are deposited along the coastline in deltas, mudflats, lagoons, and beaches. Sand-size particles tend to remain close to the shore, whereas finer silt and clay eventually settle in

---

HOW GEOLOGISTS DESCRIBE COMMON SEDIMENT PARTICLES

GRAVEL – (includes pebbles, cobbles, and boulders) – coarser than 2 mm diameter (1/10 inch)

SAND – between 2 mm and 1/16 mm (1/10 and 1/400 inch)

SILT – between 1/16 and 1/256 mm (1/400 and 1/6500 inch)

CLAY – finer than 1/256 mm (62 microns) (1/6500 inch)

MUD – a mixture of clay and silt

---

deeper, calmer water farther out at sea. There are also isolated places with abundant marine life, like tropical coastlines or inland seas, where the organisms themselves produce the sediment when their calcium-carbonate-mineral shells accumulate after they die. Visit any reef and you will find the seafloor covered with fragments of coral, coralline algae, and seashells. In this environment, seawater can be so laden with calcium carbonate that mud-size crystals spontaneously precipitate out of the water. Rocks composed of calcium-carbonate particles are called limestone. If great volumes of seawater or salty lake water evaporate, other types of crystals, such as halite (table salt) and gypsum (calcium sulfate), can accumulate and form evaporite rocks.

How Sediments Turn to Rock

The illustration of sedimentary environments (Fig 5-1) shows us the range of possible ways for sediment to accumulate. But the process of making sedimentary rock does not stop with the deposition of sediments. Conversion of loose sediment into rock requires a combination of compaction and cementation. Compaction occurs as layers of sediments accumulate and their weight pushes the sediment grains closer together. Cementation occurs as minerals precipitate between grains, a process that can occur at the same time as compaction or sometime later. Some types of sediment, such as calcite sands, can harden to rock in just a few years without deep burial, but for the majority of sediments, compaction and cementation require deep burial and long periods of time. A good place to observe this process is in the Gulf of Mexico, where oil drilling penetrates increasingly harder sediment, until it finally reaches fully hardened rock thousands of feet below the surface.

Sedimentary Rocks in the Grand Canyon and Grand Staircase

Now let's look at the variety of sedimentary rocks exposed in the Grand Canyon and Grand Staircase. As we go, keep in mind the different ways sediments accumulate on the surface of the Earth today. As we consider the evidence from the rocks, we will find remarkable similarities between deposits forming around the world today and the rock layers of the canyon.

Conglomerate

Conglomerate is a rock composed of sand or mud mixed with gravel (or larger particles). Recall that rounded stones indicate that flowing water has rolled and whittled away their sharp edges as the rocks were tumbled by streams or surf. Such stones are commonly observed in a walk along the edge of the Colorado River (Fig 5-2). In contrast, angular particles indicate relatively rapid transport and

Figure 5-2. Rounded stones along the edge of the Colorado River. *Photo by Gary Ladd.*

burial, such as in a landslide, where there is little time to smooth off the edges. There are multiple layers in the Grand Canyon that contain conglomerates with highly rounded cobbles (Fig 5-3). What does this suggest about the origin of these layers? They very likely represent the presence of ancient rivers that carried the stones from long-vanished, distant mountains.

Sandstone

Sand grains that have become cemented together are called *sandstone*. If you look at a sandstone on

Figure 5-3. Conglomerate from the Supergroup. Cobbles in the right photo eroded out of a Supergroup conglomerate and look just like river rocks! *Photos by Stephen Threlkeld (left) and Gregg Davidson (right).*

Figure 5-4. The Tapeats Sandstone, forming a cliff at the edge of the Colorado River, River Mile 117. *Photo by Tim Helble.*

Figure 5-5. Cross-bedded Coconino Sandstone. *Photo by Wayne Ranney.*

a cloudless day, you can often see the tiny mineral surfaces sparkling in the sunlight. The glassy sand grains typically are quartz crystals derived from older igneous rocks like granite, from metamorphic rocks like schist, or from older sandstones.

Sand can be transported and deposited by wind or water. Deposition in water is the most common. For example, the Tapeats Sandstone in the Grand Canyon has fossils of primitive shallow-water marine animals, indicating that these sandstone beds were deposited underwater in a shallow sea (Fig 5-4).

The Coconino Sandstone in the Grand Canyon is one of the best examples in the world of a wind-deposited sandstone (Fig 5-5). The distinctive, large-scale cross bedding in the unit has preserved the migration of giant dunes across an extensive ancient desert landscape. Evidence that these dunes formed on land includes the steepness of the preserved dune faces and the presence of land-animal tracks on many of them. The formation of cross bedding will be explained in the next chapter.

Siltstone

Silt grains cemented together are called *siltstone*. Silt deposits are commonly found in environments that lie between places where sand and clay are being deposited: on river floodplains, on tidal flats, and in shallow water offshore from river deltas (Fig 5-1). Many formations in the Grand Canyon that are typically referred to as sandstone or shale, like Tapeats Sandstone or Bright Angel Shale, actually contain many interbedded siltstone layers (Fig 5-6).

Figure 5-6. Interbedded sandstone and siltstone (thicker layers) and shale (thinner layers) in the Dox Formation, Grand Canyon Supergroup. *Photo by Gregg Davidson.*

Shale

Shale is a rock type composed of very-fine-grained clay or mud (often a mix of clay and silt). The fine-grained texture usually makes shale much softer and more crumbly than other rocks. If we find a shale layer sitting on top of a siltstone or sandstone, we know from our experience in seeing how different particle sizes settle out of water that

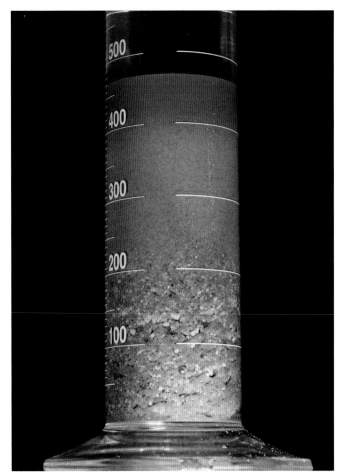

Figure 5-7. Shake up a cylinder of mixed sediments in water and let it settle. Graded bedding forms with coarse grains at the bottom and fine on top. *Photo by Joshua Olsen.*

Figure 5-8. Tonto Plateau: where the Bright Angel Shale forms a bench on top of the harder Tapeats Sandstone. *Photo by Mike Buchheit.*

Figure 5-9. Esplanade Platform: where the shale of the Hermit Formation forms a bench on top of the harder Esplanade Sandstone. *Photo by Wayne Ranney.*

the environment must have changed from steadily moving water to relatively still water – sand settles out of moving water, but clay stays suspended in water unless the flow is greatly reduced. This is demonstrated by an easy experiment that anyone can do (Fig 5-7).

When geologists find changes in particle size within or between rock layers, they immediately think about how the current's speed or depth must have changed. For example, consider what happens when a dam is built across a valley to halt a flowing river. Above the dam, sand-bar deposits from the original stream are submerged as the water begins to rise, and fine-grained silts and clays then accumulate on top of the sands. Along the coast, the same effect is accomplished by a rising sea level.

Because shale crumbles fairly easily, it tends to erode much more readily than other types of rock layers, resulting in the flattish benches seen in the Grand Canyon. There are many thin shale layers in the canyon, but the two thickest formations of shale, the Bright Angel Shale and the Hermit Formation, form the two bench areas known as the Tonto Plateau and the Esplanade Platform, respectively (Figs 5-8, 5-9).

Limestone

Limestone is rock formed from accumulations of calcium-carbonate shell fragments and/or from limey mud. Lime sediment is accumulating today across the Great Bahamas Bank, seaward of the Florida Keys and behind them in Florida Bay, along the southern coast of the Arabian (Persian) Gulf, and anywhere coral reefs are growing. How the sediment forms can be observed while snorkeling around a reef. Dead coral and algae are broken down by waves (most effectively during storms) and by tiny organisms that bore into the reef. Some fish, like parrotfish, even bite off pieces of coral and pass the sand and mud-sized particles through their guts before discharging them to the seafloor. Waves and currents carry lime sediment and the remains of sea life produced around the reef (such as primitive sea animals called crinoids) across the shore to beaches (Fig 5-10). Most limestones take a very long time to harden, but on tropical coasts

Figure 5-10. Abundant fossilized crinoid stems in limestone in the Kaibab Formation. *Photo by Wayne Ranney.*

cementation can begin to harden lime sediment into beach rock almost as soon as the sediment is buried.

Multiple limestone layers appear in the walls of the Grand Canyon. The most prominent is the Redwall Limestone, which forms the largest cliff faces in the canyon (Fig 5-11). The rock itself is not actually red. It is a gray limestone that only looks red because iron oxides derived from iron-rich minerals in the overlying Supai Group and Hermit Formation have washed down and stained the limestone surface.

Figure 5-11. Redwall Limestone at river level. *Photo by Tim Helble.*

Limestone formation is not limited to marine settings. We also find it in some lakes, particularly in arid environments. Ancient limestone layers that formed in lakes, in contrast to those formed under the ocean, can be distinguished based upon the presence of freshwater fossils. For example, much of the Claron Formation, exposed in Bryce Canyon, is known to be a lake deposit because of its abundance of freshwater-snail fossils.

No limestone has ever been documented to form from floodwater – either in the laboratory or from field observations – not even in floods as massive as those forming the Channeled Scablands in Washington State (discussed in Chapter 16). Quite simply, limestone is one type of rock that takes a long time to be deposited – much, much longer than the time span of a flood.

Evaporite Rock

High evaporation rates in open water bodies can cause mineral salts to precipitate out of the concentrated fluid, forming another type of sedimentary rock called an *evaporite*. The fluid may appear as a milky ooze on the seafloor or lake bottom, eventually hardening into crystalline rock. Thick deposits of halite (sodium chloride), gypsum (calcium sulfate), or other salts may accumulate if an inlet of the ocean is cut off from the main body of the sea, or if freshwater ponds develop in closed basins where there is no river outlet (Fig 5-1). Examples of where salts are forming today include the Great Salt Lake in Utah, the Persian Gulf, and the Dead Sea in Israel (Fig 5-12). Both the Kaibab and Toroweap Formations in the Grand Canyon contain gypsum, but it is not commonly seen in exposed rock because gypsum is easily dissolved and carried away by rain.

Figure 5-12. Evaporites forming at the edge of the Dead Sea. *Photo by Mark Wilson.*

Sequence of Layers

Now, look again at the illustration on page 55 and imagine, over time, the sea advancing over the land and later retreating. During an advance of the sea inland (or *transgression*), deeper seawater covers the mud and sand where the edge of the sea was found earlier, resulting in limestone forming over the mud and sand (the latter eventually hardening into shale and sandstone, respectively). This simple process can easily explain why the Muav Limestone covers the Bright Angel Shale, which in turn covers the Tapeats Sandstone in the Tonto Group: an ancient sea during the Cambrian Period advanced eastward over exposed Precambrian bedrock (illustrated by the upper sequence in Fig 5-13). Conversely, as the sea retreats from the land (or *regression*), one would expect sandstone to overlie limestone or shale (lower sequence in Fig 5-13). We see this within the Supai Group, where sandstones and shales of the Manakacha Formation locally overlie limestones of the Watahomigi Formation. We find the same general pattern where the Supai Group, which is predominantly layers of sandstone and shale, overlies the Redwall Limestone, and where the Coconino Sandstone overlies the Hermit Formation.

Erosion of layers is also likely to occur when sea level falls, which complicates our simple illustrated scenario. In this case, the eroded surface,

called an *unconformity*, represents missing material (gaps in the rock record). The Grand Canyon contains evidence of many sea level advances and retreats, with intermittent losses of rock to erosion. We will revisit evidence for unconformities in Grand Canyon layers in Chapter 10.

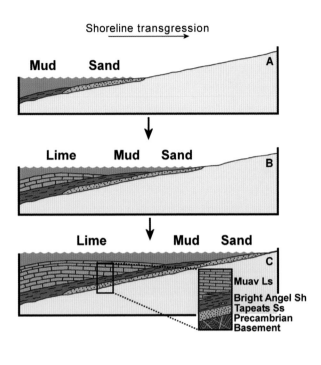

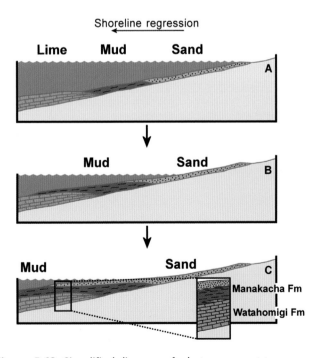

Figure 5-13. Simplified diagram of what we expect to see as sea level rises (upper sequence), or falls (lower sequence).

How Fast Do Sedimentary Sequences Form?

As we have said before, how fast a rock forms depends on its type. Volcanic eruptions can create thick ash layers in a matter of hours, while accumulations of seashells and reef colonies take decades to millennia to build up substantial layers. For some types of sediments, thin deposits can be laid down quickly, though considerable time usually passes between events. A common example is an underwater landslide. Material from a submarine landslide can create a slurry of water and sediment (called a *gravity current* or *turbidity flow*) that can travel considerable distances over the bottom and then rapidly form a new deposit (Fig 5-16, page 64). But a long time usually has to pass before another landslide adds the next layer.

Rivers today drop immense volumes of sediment into the world's oceans, producing thick sequences of sand, silt, and clay. As sea level changes, the mouths of these rivers move inland or seaward, spreading out the area where sediments are accumulating. This process, occurring over millions of years, can easily account for the sedimentary layers observed in the Grand Canyon.

A CLAY PROBLEM

Most of the grains in shale rock are composed of clay materials. Shale makes up roughly 50 percent of sedimentary rock in Earth's geologic record. That's a lot of clay! Clays are not just finely ground up sand and silt particles, they are minerals formed through chemical processes of weathering as water and air react with exposed older rocks and minerals at the surface. It takes thousands of years for a thin layer of clay-rich soil to form on igneous rock. Some flood geologists claim that great stockpiles of sediment on the Earth, over 100 million cubic miles of it, provided the grains for sedimentary rocks deposited during the flood. But considering how clays form from older minerals, how did tens of millions of cubic miles of clay minerals form in just 1650 years between the creation week and Noah's flood?

A popular flood geology explanation for the Tapeats Sandstone, Bright Angel Shale, and Muav Limestone is that advancing flood waters dropped out coarse material first (gravels and sand), followed by finer particles (silt, clay, and lime), as the flood waters deepened and slowed down, much like the lab experiment described in Fig 5-7. This might sound possible, but on closer inspection we find several inconsistencies. Alternating coarse and fine layers are found *within* these formations, burrows made by small organisms show evidence of "life as normal" (not rapidly buried by tons of sediment), lime particles are not necessarily any smaller than clay particles (so they should settle out together), and the top of the Muav, said to have formed in the deepest water, has channels quite consistent with a *tidal* environment.

Flood Geology Claims for Rapid Deposition – How Do They Stack Up?

Limestone Formation

Flood geologists speculate that ancient limestone formed when hot water that was saturated with dissolved calcite from the "fountains of the deep" encountered currents of cooler seawater loaded with lime sediment containing corals, shells, and other marine life. This combination of currents and lime sediment caused pure limestone to rapidly precipitate out of solution and settle beneath the rising floodwaters. But there are problems with these speculations. Calcite is one of the few minerals that is more soluble in *cold* water than in *hot* water, meaning that cooling would work *against* limestone formation. Also, with a violent flood, lots of clay and quartz sand should be mixed together with the lime material. Why, then, are clay and quartz sand almost entirely absent from limestone deposits like the Redwall?

Some flood geologists have speculated that the sediment making up thick limestone formations like the Muav and Redwall was derived from preexisting stockpiles of lime sediment transported from some distance to the canyon. The problem with this idea is that as a limestone deposit weathers, it simply dissolves. Broken pieces of limestone do not tend to survive transport over long distances because they dissolve away. Modern limestone deposits are clearly accumulations of calcium carbonate shell debris deposited close to where the shells formed, not carried in from distant sources of eroded limestone or from stockpiles of shells and lime mud. The idea of vast,

Figure 5-14. Staghorn coral fronds are known to grow rapidly at their tips – as much as 4 to 8 inches per year. *Photo by Tim Helble.*

Figure 5-15. Broad coral forests, like this one, grow much more slowly than staghorn coral. *Photo by Tim Helble.*

pre-flood sources of lime sediment appears to have been invented to support a particular Bible interpretation consistent with rapid accumulation of thick limestone.

There are also problems with flood geology explanations for rapid accumulation of coral-reef-type limestone *after* the flood. Citing reports of modern corals growing many inches in a single year, flood geologists argue that thick coral reefs seen around the world, some more than 1,000 feet thick, could have grown very quickly in the 4,300 years or so since the flood. But critical details are left out. For example, the fast coral growth that is observed occurs at the ends of narrow, finger-like fronds. The average rate of growth along the entire reef surface is substantially lower. (Figs 5-14, 5-15).

Figure 5-16. Illustration of a common type of gravity current, known as a turbidity flow, moving along the ocean floor. *Illustration © The Open University, all rights reserved.*

Gravity Currents

Flood geologists claim that thousands of feet of sediment could have accumulated all over the world by a rapid succession of massive submarine landslides that produced gravity currents with high concentrations of fluidized sediment (so-called *hyper-concentrated gravity currents*). There are at least two major problems with this idea. First, there is a source problem. Depositing thousands of feet of sediment by gravity currents would require hundreds or thousands of enormous stockpiles of different sediment types (and fossils), totaling millions of cubic miles, to be positioned strategically all over the world and distributed later by the flood. Many of these would have to be miraculously protected from the earlier flood violence to be available as source material for later gravity flows.

The problem is magnified when we think about how tall those stockpiles had to be. Gravity currents don't flow up hill, which means the stockpiles had to be at least *twice as high* as the thick, widespread deposits they are purported to have formed. The math is simple here: if the top half of a stockpile is swept away and deposited in a series of stacked gravity currents,

the remaining stockpile will be the same height as the deposited sediments. With no height difference, the gravity flows would then cease. So the bottom half of these stockpiles should still be around to discover and examine. The only alternative is that they were miraculously perched on top of high plateaus all over the world – just waiting for the coming flood.

The second problem is that flood geologists imply that essentially all gravity currents triggered by submarine landslides during the flood were hyper-concentrated. In reality, lower concentration

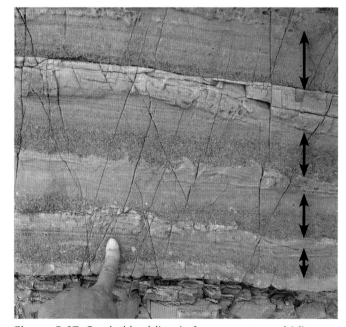

Figure 5-17. Graded bedding in four separate turbidity-current layers, identified by the arrows, with normal deposition in between. Permian rock, Inyo County, California. *Photo by Tim Cope.*

turbidity flows are the most common type of gravity current triggered by submarine landslides. Studies of turbidity flows in the lab or modern natural settings (such as California's Monterey Bay) demonstrate they produce the same characteristic grading that we saw in our sediment-settling experiment, with coarse grains at the bottom grading to finer grains toward the top (Figs 5-7, 5-17). Localized graded beds in the Grand Canyon could well have been formed by an occasional turbidity flow, but they are far from the norm there.

Migrating Dunes

Another argument for rapid accumulation is that thick, cross-bedded layers such as the Coconino Sandstone can be accounted for by plausible flow rates transporting sediments as migrating underwater sand dunes. Flood geologists claim that an underwater current moving no more than 2 to 4 miles/hour across northern Arizona could have transported and deposited the entire Coconino Sandstone, which is up to 1,000 feet thick, in a matter of days. It is true that water flowing at a rate of 4 miles/hour across a broad sandy bottom can eventually build up underwater dunes that are as high as the cross beds seen in Grand Canyon sandstones. So what information is left out? To deposit the Coconino Sandstone in a matter of days, we don't just need some sand traveling in a 4 miles/hour current, we need a *wall of sand hundreds of feet thick* sliding laterally at this speed across thousands of square miles. The migration of entire sand dunes, whether by wind or by water, is much, much slower than the movement of sand over the surface of a dune.

Preliminary Comparisons: How Plausible Is the Flood Geology Model?

Although we are far from being ready for a full comparison of the conventional and flood geology models, much can be learned from a simple understanding of the types of sedimentary rocks and how they form. The sedimentary layers found in the Grand Canyon can be easily explained by a succession of rising and falling sea levels. No fantastic or undiscovered natural processes need be invoked to account for what is observed. The flood geology model, on the other hand, requires many fantastic or never-before-seen explanations, including sediments accumulating at phenomenally high rates, unreproducible chemical reactions occurring in deep ocean fissures, mysterious lack of mixing of clay and lime in the same layers, monumental stockpiles of pre-flood sediments awaiting redistribution, and walls of sediment hundreds of feet high moving as a unit across continents. It's remarkable that such speculations are even necessary, given the total absence of any descriptions of global tsunamis, catastrophic continental upheavals, massive gravity flows, or violations of natural laws in the Genesis account of Noah's flood.

FISH THAT DIED WHILE EATING: EVIDENCE OF RAPID DEPOSITION BY A GLOBAL FLOOD?

Flood geologists routinely show images of fossils of fish buried and preserved in the act of eating another fish as evidence of a very sudden and massive deposition of sediment from a global flood (Fig 5-18). But lots of information is missing. For one, there are multiple ways fish can suddenly be killed (even in the midst of a meal), such as fallout from a volcanic eruption, or drifting into water with insufficient oxygen. What's more, these "fish-eating fish" fossils were found in layers that many flood geologists say are post-flood deposits (Cenozoic). This means a *"post-flood"* fossil is used as support for a global flood!

Figure 5-18. Fossil of fish eating another fish on display at Fossil Butte National Monument, Wyoming. *Photo by Tim Helble.*

Mud cracks peeling up after desiccation. *Photo by Bronze Black.*

CHAPTER 6

SEDIMENTARY STRUCTURES
CLUES FROM THE SCENE OF THE CRIME

by **Carol Hill** and **Stephen Moshier**

WHEN SEDIMENTS ARE DEPOSITED, they leave clues about the conditions of their deposition. *Sedimentary structures* are one of those clues (other clues include particle size, fossils, rock chemistry, and even simple things like color). Sedimentary structures usually form at the same time as, or just after, the deposition of sediments. Like footprints and other evidence at the scene of a crime, sedimentary structures can tell geologists what happened, even if no one was there to see it. For example, when wet mud on a modern tidal flat or flood plain dries under the sun, the mud shrinks, cracks, and curls, forming structures called *mud cracks*. So if we see mud cracks in an ancient rock, we can surmise that they formed in the same manner as today – by the drying and cracking of wet mud.

Sedimentary structures are abundant in Grand Canyon rock, and they illustrate with amazing detail the environments and conditions of their deposition. None of the sedimentary structures in the Grand Canyon require explanations of unusual or abnormal circumstances, such as a global deluge that would have happened only once in Earth history. In fact, most of the structures in Grand Canyon strata are commonly associated with modern shallow-marine to coastal-land environments. In this chapter we will compare modern (recently formed) structures, like

Figure 6-1. Mud cracks formed recently in wet mud along the Little Colorado River. *Photo by Bob Buecher.*

raindrop prints, ripple marks, mud cracks, and cross bedding, with similar structures in ancient Grand Canyon strata.

Mud Cracks

Since we have already mentioned them, let's start with mud cracks, which are also called "desiccation cracks." Invariably, the presence of mud cracks implies mud baking under the sun (that is, the cracks form above water). Preservation occurs when sediment of a different texture or color gets washed into the cracks by the next storm or flood, or by minerals that fill the cracks much later in time. Mud cracks that formed recently, after high water receded along the Little Colorado River, look remarkably similar to ancient mud cracks formed in the 525-million-year-old Tapeats Sandstone (Figs 6-1, 6-2). Desiccation cracks with polygons 6 inches or more in diameter can be seen in the Coconino Sandstone (Fig 6-3).

CAN MUD CRACKS FORM IN A RAGING FLOOD?

Flood geologists have argued that features similar to mud cracks can form on the sea floor in deep water as fine-grained sediment settles and "dewaters." While in some cases these deep-water, or *syneresis,* cracks can resemble polygonal mud cracks, more often they are spindle-shaped or sinuous – having no continuous pattern and not connecting to form geometric shapes (Fig 6-4). Flood geologists quote the mainstream geology literature on syneresis cracks when they interpret what they observe in the Grand Canyon, but the descriptions and photographs of syneresis features in those scientific papers *do not* resemble the mud cracks in the Grand Canyon strata. True mud or desiccation cracks, which are diagnostic of *above-water* exposure, are further distinguished from deep-water cracks by the presence of raindrop prints, ripple marks, and vertebrate tracks in the same rock – such as is the case for the Coconino Sandstone in the Grand Canyon.

Figure 6-2. Mud cracks in the Tapeats Sandstone, filled in with calcite. *Photo by Doug Powell.*

Figure 6-3. Polygonal desiccation cracks in the Coconino Sandstone. *Photo by David Elliott.*

Figure 6-4. Spindle-shaped syneresis cracks in the Pennsylvanian-age Stellarton Formation, Nova Scotia. *Photo by M.C. Rygel.*

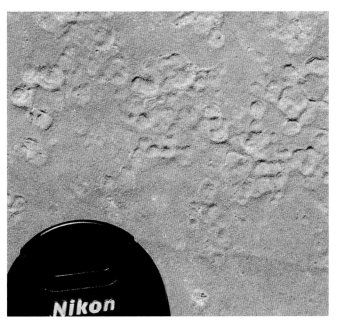

Figure 6-5. Fossil raindrop prints in the Coconino Sandstone. *Photo by David Elliott.*

Figure 6-6. Ripple marks (and other interesting things) in about 20 feet of water, formed by a current of about 0.5 to 1.0 mph. *Photo by Grant Johnson.*

Raindrop Prints

Raindrop prints are made when droplets of pounding rain impact wet mud, silt, or sand, thus creating small depression imprints of those drops in the sediment. This can only happen when moist sediment is exposed to the air, because if the sediment is underwater it cannot be impacted by raindrops. In other words, this feature *could not* have formed in a rapidly rising floodwater environment or even in a body of water deeper than a few inches. Raindrop prints have been found in the Coconino Sandstone at many locations (Fig 6-5).

Ripple Marks

Ripple marks are typically generated by currents moving in one direction (Fig 6-6) or by the to-and-fro motion of waves in shallow water (generally not more than a few tens of feet deep). Ripple marks that formed along the bank of the Colorado River in the Grand Canyon during a high flow event in 2004 are very similar to ripple marks that formed in the Tapeats Sandstone 525 million years ago (Figs 6-7, 6-8). Ripples have also been photographed on the sea floor in very deep water, where slow-moving density-driven currents cause these features. However, high-velocity currents produced by tsunamis,

Figure 6-7. Ripple marks preserved in the Tapeats Sandstone. *Photo by Alan Hill.*

Figure 6-8. Ripples formed on a sand bar deposited by the river at high water. *Photo by David Rubin.*

such as proposed by flood geologists for a global flood, churn up bottom sediments and do not produce regularly spaced sets of ripples like the ones shown in the photos on the previous page.

Cross Bedding

As was mentioned in Chapter 5, cross bedding is a feature commonly observed where currents (water or wind) transport and deposit sand in large ripples or dunes. Sand grains are swept along by water or wind currents up the more gently sloping back side of a ripple or dune, and are deposited on the steeper

Figure 6-11. Avalanche deposits on a crest of the Kelso dunes, Mojave National Preserve, California. Avalanching occurs when the slope angle reaches 30-34°. *Photo by Mark A. Wilson.*

"downstream" side. The accumulating sand on the advancing front of the dune is sorted by size and density into a series of very thin layers that are inclined relative to the sand layer as a whole. As the crest of a large sand dune moves across the desert, it leaves behind a trailing deposit of sand that can be up to 30 feet thick. The inclined layers or cross beds of the preserved layers in the trailing bed show the direction that the dune was moving, thus giving geologists information on ancient wind or water currents. Cross bedding is observed in some beds of almost every sedimentary formation in the Grand Canyon.

The largest and most spectacular cross bedding in the Grand Canyon is evident in the Coconino Sandstone, and to a lesser extent in the beds of the Supai Group (Figs 6-9, 6-10). The maximum stable angle for loose, dry sand falls between about 30 to 34°. Steeper slopes tend to fail via small avalanches in sand dunes (Fig 6-11). A bit of moisture makes sand more stable, but anyone building a sand castle on the beach knows that if the sand becomes saturated (such as being submerged at high tide), the whole structure will collapse. Saturated sand in underwater dunes cannot maintain slopes as steep as dry sand in desert dunes. Cross beds in the Coconino have angles typical of desert dunes, with maximum angles of 29 to 31°.

Figure 6-9. Cross bedding in the Coconino Sandstone, Bright Angel Trail. The angle of cross bedding is about 30°. *Photo by Tim Martin.*

Figure 6-10. Cross bedding in sandstone of the Supai Group, with a slope of about 20°. *Photo by Tim Helble.*

Tracks

Preservation of a mark or disturbance made in sediment by an organism is referred to as a *trace fossil* – meaning traces were left behind of animal activity. Like mud cracks, tracks on sediment can be preserved if filled and covered by different sediment. If nothing else, tracks clearly indicate the presence of living and active creatures. Tracks of reptilian-like and scorpion-like creatures that were very much alive when they made them are found in the Coconino Sandstone on the surfaces of exposed cross bedding (Fig 6-12).

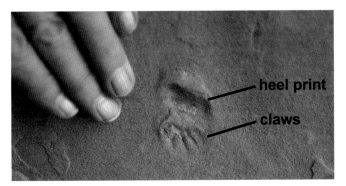

Figure 6-12. Close-up of a footprint (track) made by a small reptile in sand that later became part of the Coconino Sandstone. Note the tiny, delicate claw marks. *Photo by Cyndi Mosch.*

These gently formed and exquisitely preserved tracks are extremely difficult to reconcile with flood geology. In Chapters 4 and 5 we discussed the belief of flood geologists that the Coconino Sandstone was deposited by swift currents moving large volumes of sand rapidly across the continent. How could small animals trek across these dunes (and have their tracks preserved) in conditions that can be compared to repeated global tsunamis and giant underwater landslides? We'll discuss tracks in more detail in Chapter 15.

Conclusions

The sedimentary structures described here are not rare or isolated features in the Grand Canyon. They are present and abundant not just in the 4,000 feet of Paleozoic sedimentary formations exposed in the steep canyon walls above the Colorado River, but also in the additional 12,000 feet of underlying Grand Canyon Supergroup rock (Fig 6-13). For ex-

ample, consider the presence of mud cracks in five different Grand Canyon formations, with repeated occurrences within each formation. To account for these in the flood model, either large regions of the planet had to repeatedly flood and dry out in the midst of a year-long global flood, or some yet-to-be-identified process must have repeatedly formed cracks in deep water that mysteriously mimic what we see forming on the land today.

Why do the Grand Canyon's sedimentary structures look just like structures forming in modern environments today? What are desert sand dunes displaying animal tracks doing in the middle of a miles-deep flood? And how could delicate features such as claw marks be preserved under raging flood conditions? Sedimentary structures in strata of the Grand Canyon are powerful evidence against a flood geology origin for these rocks.

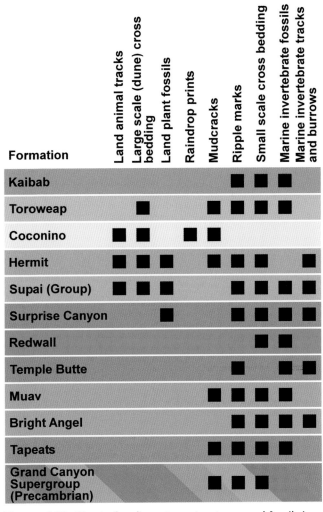

Formation	Land animal tracks	Large scale (dune) cross bedding	Land plant fossils	Raindrop prints	Mudcracks	Ripple marks	Small scale cross bedding	Marine invertebrate fossils	Marine invertebrate tracks and burrows
Kaibab						■	■	■	
Toroweap		■				■	■	■	
Coconino	■	■		■	■				
Hermit	■	■	■		■	■	■		■
Supai (Group)	■	■	■			■	■	■	■
Surprise Canyon			■			■	■	■	■
Redwall							■	■	
Temple Butte						■		■	■
Muav						■	■	■	■
Bright Angel						■	■	■	■
Tapeats						■	■	■	■
Grand Canyon Supergroup (Precambrian)						■	■	■	

Figure 6-13. Chart of sedimentary structures and fossils in Grand Canyon strata.

Recent rockfall from the Redwall Limestone in Marble Canyon about River Mile 40. This exposes the true color of the Redwall Limestone. *Photo by Tim Helble.*

CHAPTER 7

USING THE PRESENT TO UNDERSTAND THE PAST

by **Stephen Moshier** and **Gregg Davidson**

UNVEILING THE PAST STARTS WITH an observation that processes at work today are producing rock and sediment formations with characteristics that are readily seen in ancient rocks. In the previous two chapters we saw that ancient sedimentary rocks, such as the Grand Canyon strata, can be compared to sediments accumulating on the land and in the sea today. The particles and sedimentary structures in ancient conglomerates, sandstones, shales, and limestones are not fundamentally different from sediments we find associated with modern rivers, lakes, bays, and the sea.

After 200 years of geologic study, it is clear that the basic geologic processes creating rocks today have been active over much of the history of planet Earth. These observations have led to an appreciation that the same physical and chemical laws in effect today were also in effect in the past, allowing us to use modern observations to identify ancient events and environments. This *uniformity* of natural laws gave rise to the term *uniformitarianism* and to the *Principle of Uniformitarianism,* which is one of the fundamental precepts of geology. Geologists have long recognized that natural processes can be slow (like coral reef growth or deep ocean sedimentation) or fast (like volcanic ash deposition), though some geologists of the nineteenth century tended

to think that most of the Earth's rocks represented slow processes. Today, uniformitarian geology recognizes that there are many places and times where catastrophic events have contributed to shaping the Earth's varied layers, and that the physical conditions on Earth, such as the chemical makeup of the atmosphere and oceans, have not always been the same as they are today.

Misrepresenting Uniformitarianism

Flood geologists commonly demonize uniformitarianism by misrepresenting it as being synonymous with *materialism* or *evolutionism.* Yet when they seek to find scientific evidence in support of a young Earth, they actually apply uniformitarian principles! For example, some prominent flood geologists believe that the 1980 catastrophic eruption of Mount St. Helens provides clues to a rapid formation of the Grand Canyon. Specifically, the volcanic eruption delivered dozens of feet of sediment and ash in the valley below the mountain in the span of just hours. Later, an impressive gorge was carved in the very-soft ash sediment. The deposits contain layers and bedding, which flood geologists say look like rocks and cliff faces in the Grand Canyon.

This comparison of modern deposits to ancient deposits is a fully uniformitarian exercise – the present (Mount St. Helens) is the key to the past (Grand Canyon). In this case, however, as in many other examples, flood geologists are not consistent in comparing apples with apples and oranges with oranges. The formation and behavior of volcanic ash layers cannot be directly compared with the formation and behavior of sandstone or limestone. The rocks in the Grand Canyon are made of entirely different materials from ash, and the immense size of the canyon's vertical cliffs (*much* larger than those found at Mount St. Helens) clearly testifies to the fact that the cliffs were already hardened to rock before they were cut by the Colorado River.

Flood geologists further depart from uniformitarian principles (and from Christian doctrines of God's consistency and providence) when they assume that natural laws describing physical and chemical processes must have been different during the creation week, before the fall in the Garden of Eden, or at various points during Noah's flood. Some Young Earth advocates write that the natural laws in the whole universe are a consequence of God's "curse" or punishment for the fall. These arguments fail to be supported by either science or Scripture. In science, all observations point to a consistency in the laws of nature, back to the first microseconds of the universe. The Bible likewise says nothing about the fundamental laws of nature being altered after man's sin.

Using Present-Day Landscapes to Recognize Ancient Landscapes

If the present is the key to the past, can we identify landscapes today that are comparable to Grand Canyon landscapes that existed in the past? We can indeed. To illustrate, we will visit five different sites along the canyon walls, each representing a different ancient landscape. We will apply what we described in the previous two chapters about sedimentary rocks – their composition and structures – to interpret the past.

Landscape 1: *Bare Naked Rock (Crystalline Basement Rock Exposed by Erosion)*

Our starting point, at the bottom of the canyon, is the Great Unconformity carved into the ancient Vishnu Schist (Fig 7-1). A modern analog (meaning a site today with very similar characteristics) is found in Canada, north of the Great Lakes, where large tracts of land are worn down to a nearly flat surface, exposing some of the oldest rocks on Earth (over 4 billion years old) (Fig 7-2). If today's ocean were to advance over this crystalline rock and begin to deposit sediment upon it, that would be comparable to the sediments of the Unkar Group being deposited over the older, eroded crystalline rock in the eastern part of the Grand Canyon.

Figure 7-1. The horizontally layered Tapeats Sandstone over the Vishnu Schist along the Great Unconformity (hand pointing to the Great Unconformity). *Large photo by Wayne Ranney, inset photo by Gerry Stirewalt.*

Figure 7-2. Landscape #1. Ancient banded metamorphic rock eroded nearly flat along Georgian Bay, Lake Huron, Ontario. *Photo by Ted John Jacobs.*

Figure 7-3. Bright Angel Shale wall in the eastern Grand Canyon. *Photo by Tim Helble.*

Figure 7-4. Landscape #2. The reddish-brown muds of the Rio de la Plata, Argentina. *Photo by Wayne Ranney.*

Landscape 2: *Muck and Mud (Clay Deposition in a Shallow Near-Shore Sea)*

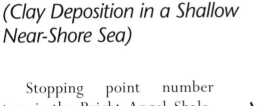

Stopping point number two is the Bright Angel Shale, measuring about 270 feet thick in the center of the canyon (Fig 7-3). The clay-rich sediment of the Bright Angel Shale contains fossils of trilo-bites and abundant brachiopods, as well as fossil tracks and burrows.

Though these particular creatures are no longer with us, we nonetheless find modern environments that are accumulating the same-size particles (mud), where bottom-dwelling marine organisms make similar tracks and burrows. The muddy sea floor beyond the mouth of the Rio de la Plata in Argentina provides a modern example of the depositional setting for the Bright Angel Shale (Fig 7-4).

Landscape 3: *Vacation Destination (Warm Seas and Carbonate Deposition)*

Moving upslope in the Grand Canyon to the Muav Limestone, we find carbonate sediments made from shell material, coral, and limey mud deposits (Fig 7-5). The transition from shale to carbonate here is consistent with rising sea levels (or subsiding land levels) and a change in the local environment to a more offshore setting that became isolated from the influx of clay. Modern analogs are found in places like the southern coast of the Arabian Gulf (Fig 7-6).

Figure 7-5. Muav Limestone wall, between the multi-hued brown layers of the Bright Angel Shale below and the Redwall Limestone above. *Photo by Tim Helble.*

Figure 7-6. Landscape #3. Abu Dhabi Coast at Khor al Bazam, United Arab Emirates. Light blue in the right-hand image is shallow water where carbonate deposition (limestone) is occurring. *Photos: MODIS Rapid Response Team, NASA/GSFC; Small inset photo by Tim Helble.*

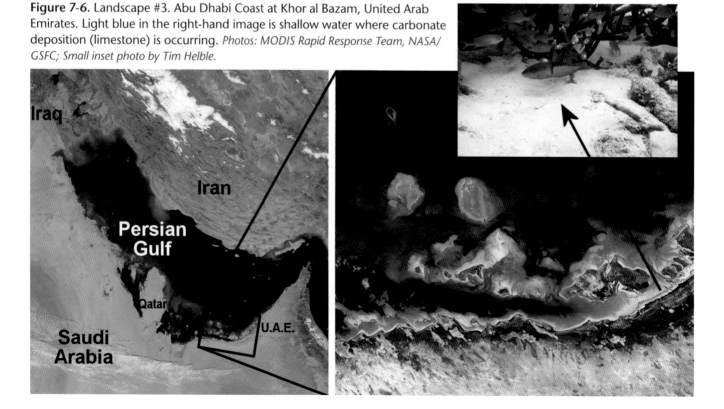

Landscape 4: *Subterranean Labyrinth (Cave and Sinkhole Formation)*

Stop number four is the Redwall Limestone, one of the most prominent rock units within the Grand Canyon (Fig 7-7). It is the red unit that forms the sheer cliff face, from 500 to 800 feet thick, about halfway down from the canyon rim. If you 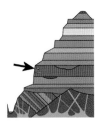 look closely at the top of the Redwall cliff, you may see a horizontal layer with lots of holes in it. Some of these holes are entrances to caves that formed less than 10 million years ago. But much, much earlier than that, about 320 million years ago, the then-recently deposited Redwall Limestone was uplifted slightly above sea level. That uplift allowed freshwater to descend and circulate through the limestone, pockmarking it with thousands of sinkholes and caves over much of the area that is now northern Arizona. Caves commonly form in limestone when it is exposed to naturally acidic

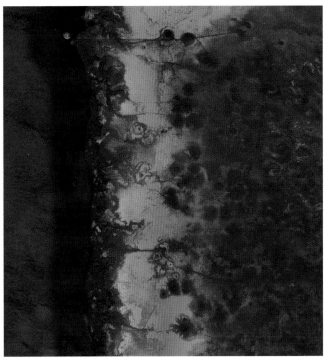

Figure 7-8. Landscape #4. Aerial photo of the Caribbean coast of Yucatan, showing cenotes, or sinkholes (the dark round circles), in the gray limestone. *Copyright 2013 TerraMetrics, Inc., www.terrametrics.com.*

Figure 7-7. Caves in the Redwall Limestone, eastern Grand Canyon. *Photo by Bronze Black.*

Figure 7-9. Cenote (sinkhole) at Chichén Itza, Yucatan, Mexico, located only a few hundred feet above sea level. *Photo by Emil Kehnel.*

Figure 7-10. A scuba diver near the bottom of a deep cenote in the Yucatan Peninsula, Mexico. *Photo by HP Hartmann.*

rainwater (conditions that exist above sea level). Caves near the surface often eventually collapse, resulting in sinkholes forming on the surface and caves filling with overlying material. (We will cover the topic of these old "paleo" caves in Chapter 10.)

We see this process occurring today in places like the Yucatan Peninsula of Mexico, where sea level has dropped and the limestone is exposed to rainwater that is slowly dissolving the rock to form caves (Figs 7-8, 7-9). Where cave roofs have collapsed, cenotes (sinkholes) have formed, and the caves are slowly filling with material washed in from the surface. In the Yucatan, these sinkholes are connected by an extensive underwater cave system that can be explored only by highly trained divers (Fig 7-10).

Landscape 5: *Hot and Dry (Desert Sands)*

The final stop is the Coconino Sandstone – the thick, distinctive, buff to light yellow layer just below the rim of the canyon (Fig 7-11). As we discussed in the previous two chapters, sand texture, cross bedding, and animal tracks in the Coconino 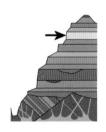 are consistent with what we find in sand dunes formed through wind-driven processes, such as those in the Namib Desert in West Africa today (Figs 7-12, 7-13, 7-14). If sea level someday rose to submerge the sand of the Namib Desert, we would expect to find shale

Figure 7-11. The Coconino Sandstone, as it appears near the Bright Angel Trail. *Photo by Wayne Ranney.*

Figure 7-12. Wind-driven depositional processes at work. *Photo by Brennan Jordan.*

Figure 7-13. Cross beds in a modern sand dune. *Photo by Marli Miller.*

Figure 7-14. Landscape #5. Shoreline along the Namib Desert, West Africa. *Photo by Amy Schoeman.*

or limestone deposits beginning to form above the sand, just as we find the Toroweap (siltstone) and Kaibab (limestone) formations sitting on top of the Coconino Sandstone in the Grand Canyon.

Conclusion

In this chapter we visited sedimentary formations that represent five different ancient landscapes, each with analogous modern environments that are producing deposits with remarkably similar characteristics. These and other landscapes are reflected repeatedly in layers within the Grand Canyon and in the Grand Staircase to the north. As we have seen in previous chapters, no fantastic or never-before-observed mechanisms are required to account for any of these sedimentary layers.

To this point, we have noted that some rock types form quickly, like ash deposits, and others very slowly, like limestone, but we would prefer to be able to say more than just "fast" or "slow." How do geologists determine the time that was required to form an ancient rock layer and how long ago it formed? And are there debatable assumptions that have to be made to calculate ages? This brings us to the next two chapters, which address time and the ages of rocks.

The Grand Canyon Corridor from Granite Rapid to Hermit Rapid. *Photo by Bronze Black.*

TIME

The previous three chapters focused exclusively on what can be learned from sedimentary rocks. As we address the subject of time, we will broaden the scope of our discussion to include all rock types and any features found in them. Here we are ready to ask how geologists determine the age of various rocks, how a sequence of events can be determined from clues left behind in the rocks, and how assumptions used in calculations can be tested to determine if they are valid. For this subject, we'll start Chapter 8 where the first geologists did, by making some logical observations to determine which rock units or features must be older and which must be younger, initially without any knowledge of their actual age.

CHAPTER 8

SOLVING PUZZLES
RELATIVE DATING AND THE GEOLOGIC COLUMN

by **Stephen Moshier** and **Gregg Davidson**

I**N THE SEDIMENTARY ROCK** chapters, we focused on recognizing processes and environments in individual layers of rock — essentially, on the history of one rock unit. We are now ready to expand our studies and try to piece together a larger history that incorporates observations from multiple rock layers. In any history, an essential step is to determine the order in which events occurred. At this stage, we are not thinking about assigning a particular age to any layer or feature — we just want to know what happened first, then second, etc. Initially, this may seem terribly simple. Don't we just start counting at the bottom and work our way up? But what do we do with tilted sedimentary layers on the right side of a trail and metamorphic rocks on

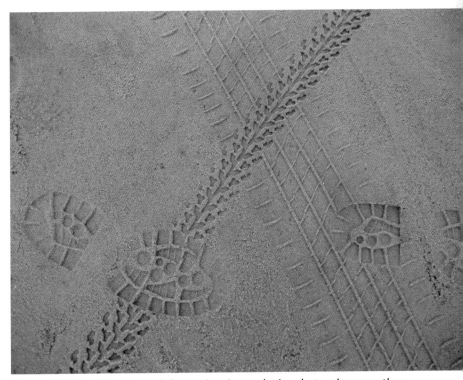

Figure 8-1. Car, bicycle, and shoe prints in sand – in what order were they made?

the left side, or with fingers of igneous rock inserted into some layers but not others? Any close inspection reveals complexity that requires careful attention to sort out.

Relative Dating

We need a set of rules for putting all the pieces together in the correct sequence to solve the puzzle. This brings us to a series of principles used by geologists for *relative dating*, where we can say "relative to that rock, this rock is older." The principles that follow are very similar to methods used in solving crimes. For example, let's say detectives have evidence that the last person to leave a property committed the crime. Joe rode his bike, Sally drove her car, and Dave walked (Fig 8-1). By looking at which set of tracks lies on top of the others, we can see that Sally left first (car tire tracks), followed by Joe (bike track), and finally by our criminal, Dave (footprints). Case solved, even though no one was there to witness the people leaving!

Many of the principles for determining order in geologic studies were worked out in the seventeenth century by a Danish natural philosopher named Nicholas Steno. A British surveyor, William Smith, contributed more principles in the early nineteenth century. To illustrate several of their principles, let's look at a simplified cross section above. (We'll make repeated reference to this example cross section as we describe the various principles; Fig 8-2.)

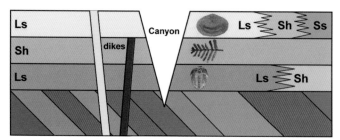

Figure 8-2. Hypothetical geologic cross section, illustrating the principles of relative dating (not a depiction of the actual Grand Canyon). Ls = limestone, Sh = shale, Ss = sandstone.

Principle of Superposition

The first principle is about the order in which sedimentary layers accumulate. In any sequence, each layer had to be present before the next could form. If you pile a stack of books on your desk, which book did you put down first, the book on the top or the book on the bottom? The simplest answer is that the book on the bottom was put down first. Assuming no extraordinary disturbance, sedimentary layers follow a simple sequence of superposition – of younger placed on top of older – as seen in the top three layers of our example cross section (Fig 8-2).

We always have to pay attention to potential disturbances, however. Faulting, for example, can push

Figure 8-3. According to the Principle of Superposition, this sequence of layers near the rim of the canyon, from oldest to youngest, is the Hermit Formation, Coconino Sandstone, Toroweap Formation, and Kaibab Formation. *Photo by Tim Helble.*

Kaibab Formation

Toroweap Formation

Coconino Sandstone

Hermit Formation

older layers on one side higher than younger ones on the other side. In our book example, if we find all three books torn in half, we have reason to believe that the original order may have been altered. We also need to be reminded that superposition applies only to sedimentary layers. An igneous intrusion inserts younger rock *within* older rock (so older rock is both below and above the younger igneous rock).

The idea of choosing interpretations that are the simplest or most logical is an application of Occam's Razor, which is a principle urging one to select, from among competing theories or explanations, the interpretation that makes the fewest assumptions or that requires the fewest improbable events. It is sometimes worded as "the idea of least astonishment." Modern geologists make routine use of this principle. Flood geologists claim to apply the same principle, but the explanations they offer are rarely the least astonishing. A whole series of improbable events is generally required in order to make the data fit their preconceived answer.

LEAST ASTONISHING EXPLANATION?

In the Grand Canyon, we find many discrete sediment layers with interconnected, polygon-shaped cracks that look just like modern mud cracks. The simplest explanation of the observed data is that the accumulating sediments occasionally dried out before the next layer was deposited. Flood geologists don't start with the data, however, they start with the answer – all was deposited by a global flood. To make the data fit the answer, the cracks must be deepwater compression cracks, with as-yet-unidentified processes at work to coincidentally make them look like mud cracks. Add to this some critical information left out, such as raindrop impressions and reptile tracks in some of the same layers, and the flood geology explanation is clearly not the least astonishing.

Principle of Lateral Continuity

This principle again applies only to sedimentary layers, which are often deposited in laterally extensive beds. When we find places like the Grand Canyon, with layers that can be traced from one side of the canyon to the other, it is evident that those layers were once continuous before they were separated by incision of the canyon (Fig 8-4). In other words, the

Figure 8-4. *The Principle of Lateral Continuity,* illustrated in the Grand Canyon. The layers across the canyon, from the South Rim (left) to the North Rim (right), were continuous before they were cut by the canyon. *Photo by David Edwards.*

layers did not form on opposite sides of a pre-existing canyon. Rather, the layers were deposited first, and a canyon was eroded down into the layers at some later time.

Here in the Grand Canyon, it may seem obvious that the layers were once continuous, but tracing layers across vast landscapes can sometimes be difficult.

For example, what if a layer gradually changes from limestone to shale, like in the lowest horizontal layer in our example cross section (Fig 8-2, page 82). We might be lucky enough to find a place where we can see the change from limestone to shale. If not, sometimes fossils can help us recognize discontinuous outcrops of the same layer. This brings us to another principle – one that is based on fossils.

Principle of Faunal Succession

Geologists such as William Smith noticed two centuries ago that different kinds of fossils consistently appear in the same sequence in different locations. This discovery came far in advance of Darwin's publications, and was not based on any evolutionary assumptions. The observation was not initially of increasing complexity of organisms over time, but instead was of simple replacement, on a global scale, of one group of organisms by another. The most frequently observed replacements were associations of marine invertebrates (trilobites, clams, snails, etc.) that were always found stacked in the same order. So, if we find a particular group of identical fossils in two distant rock outcrops, it is likely that those organisms were deposited in the same time frame (this example combines the *Principles of Lateral Continuity* and *Faunal Succession*).

Principle of Original Horizontality

Sediments depositing on a steep slope are unstable. They readily fail and slide or flow downslope, accumulating in horizontal beds at the bottom. Steeply sloped cross-beds can form within a single dune, but even dunes, as a whole, tend to migrate and deposit sediments on a roughly horizontal surface. So when we find steeply tilted sedimentary strata, we know they first must have been deposited on a relatively flat surface (or gentle slope) and then hardened before uplift and tilting began. Applying the *Principle of Original Horizontality* to the cross section on page 82, we know the steeply tilted sedimentary layers beneath the canyon were not deposited in that position, but were tilted sometime after the sediments had been hardened into rock (cemented).

Principle of Cross-Cutting Relationships

This principle applies to all rock types. Faults (shifted rock on either side of a break) or dikes (magma intruded into cracks) must always be younger than the layers or features they cross (Fig 8-5). A fault or dike cannot form in something that is not already there. This is a simple principle to understand. If you draw a line across a piece of paper, which is younger, the paper or the line? The most logical answer is that the line must be younger than the paper (again using Occam's razor, or the idea of least astonishment). In our example cross section (page 82), the yellow dike

Figure 8-5. A dike cutting across the Hakatai Shale of the Grand Canyon Supergroup at Hance Rapids. So which is older, the dike or the shale? *Photo by Tim Helble.*

cuts through all other layers, so it must be the youngest feature. The upper layer truncates (cuts off) the red dike, which means the surface here was partially eroded. Both the erosional event and the upper layer are younger than the red dike. Finally, the red dike cuts across all but the upper layer, making it the fourth youngest feature.

Two relative time markers that fit with the *Principle of Cross-Cutting Relationships* are contacts and unconformities. A distinct change from one rock type to another creates a line (or a plane, considering all three dimensions) known as a *contact*. This term is somewhat self-descriptive, in that it represents the surface where one layer contacts another. In the example cross section, the line between the upper horizontal layers and the lower tilted layers is a special kind of contact called an *unconformity* – a contact separating older from younger rocks and representing a gap in the geologic record. In this example, sediment is missing (and thus a gap in represented time) due to partial erosion of the top of the tilted layers before deposition of the first horizontal layer. A layer cannot be eroded if it is not already there, so the erosional surface (the unconformity) is younger than the layer itself.

Recognizing unconformities in the field is important because they indicate a significant disruption in the sequence of formation or deposition of rock units. This disruption can represent large or small gaps in time between rock layers. Multiple unconformities in a sequence of rocks, such as in the Grand Canyon, are impossible to reconcile with a single catastrophic event, which is why flood geologists work so hard at discounting the presence of all but the most obvious ones. Unconformities will be covered in more detail in Chapter 10.

We'll now make use of these five principles to solve the sequence puzzle in our example cross section (Fig 8-2, repeated below).

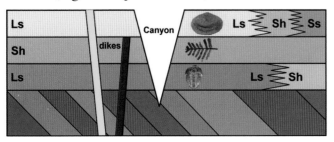

1) The layers beneath the canyon were deposited horizontally and hardened.

2) These hardened layers were deformed and tilted upward.

3) The tilted and uplifted layers were eroded to a nearly flat surface, and later buried by the first two horizontal layers.

4) Magma, forming the red dike, intruded into all the strata, and perhaps higher ones that later eroded away.

5) Erosion removed all rock higher than the top of the second horizontal layer (known because the top of the red dike is cut off).

6) The top horizontal layer was deposited.

7) Magma again intruded from below and formed the yellow dike.

8) The canyon was carved.

The Geologic Column

The principles of relative dating described above have been applied over the last two centuries at many locations around the world, in order to work out the sequence of events at each site. But how can we tell if event 3 at site A is younger or older than event 7 at distant site B? This is where fossils have played an important role.

DOESN'T DATING ROCKS WITH FOSSILS INVOLVE CIRCULAR REASONING?

Critics of modern geology claim that geologists pick dates they want for formations based upon circular reasoning that goes like this: "Geologists date the fossils by the rocks they are in, and they date the rocks by the fossils that are in them!" But this charge is wholly unfounded. Fossils were initially used to identify the different time periods in Earth's history, based on the consistent order in the fossil sequence around the world. Radiometric dates were later measured independently on rocks identified with each time period. The radiometric dates supply the age in years and verify that the relative order of fossils and geologic periods is accurate.

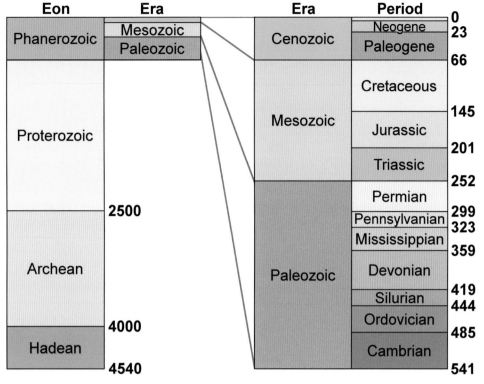

Eon	Era	Era	Period	
Phanerozoic	Mesozoic	Cenozoic	Neogene	0
				23
	Paleozoic		Paleogene	
				66
Proterozoic		Mesozoic	Cretaceous	
				145
			Jurassic	
				201
			Triassic	
				252
		Paleozoic	Permian	
	2500			299
			Pennsylvanian	323
Archean			Mississippian	
				359
			Devonian	
				419
			Silurian	444
	4000		Ordovician	
				485
Hadean			Cambrian	
	4540			541

Figure 8-6. Geologic time and the geologic column. The absolute ages (numbers in bold) are in millions of years. The order of the rock periods through time is based on relative dating methods applied over the last 200 years; the absolute ages are based on radiometric dating techniques.

Recall our discussion of faunal succession. Long before Darwin, natural scientists noticed a consistent sequence of fossil groups in sedimentary rocks. It was not initially a sense of one variety turning into another, but instead was a simple replacement of one set of organisms by another. Critically, the order of those fossil groups in different layers appeared to be consistent from one locality to the next, *even on different continents!* The catastrophists of the nineteenth century took this as evidence of a series of global catastrophic events that wiped out life forms present at the time, which then were replaced by a new set. Noah's flood was thought by some to represent the most recent catastrophe. Catastrophists and gradualists alike recognized that each of these ubiquitous fossil groups represented organisms that lived at a particular time in the Earth's history. This understanding allowed Earth scientists to extend relative-dating principles to be able to connect the sequence of events at one site to the sequence of events at a distant site. Layers at two sites with the same suite of fossils represent the same time interval. By comparing multiple sites that have overlapping layers with distinctive fossil suites, a composite history was compiled, now referred to as the *geologic column* (Fig 8-6).

Radioactivity had not been discovered when much of this work was happening, so no absolute dates or ages were assigned to the different time periods. Names were given to different rock units, usually based on the area in which the most intensive studies had been undertaken. For example, the Jurassic period is named after rocks of this age studied in the Jura Mountains of Europe. Transition from one period to the next was based on a noticeable change in the types of fossils, extinction events, the beginning of mountain-building phases, or distinctive changes in rock properties. A hierarchical time scale was developed that is analogous to centuries, years, months, days, and hours, where each unit is a subdivision of the previous, larger unit. For geologists, the units are Eon, Era, Period, Epoch, and Age (Fig 8-7). In this book, we will limit our descriptions mostly to Eras and Periods.

Prior to the discovery of radioactivity, one could speak of the Mesozoic Era as being divided up into three sequential Periods (Triassic, Jurassic, and Cretaceous), but with no knowledge of how long each period or era lasted or if the periods represented equal or different time spans. Radiometric dating finally made such determinations possible, so we now recognize that some Eras and Periods represent longer time intervals than others. More importantly, radiometric dating allowed a test of whether the relative dating methods produced an accurate composite history. If the history is correct, rocks identified as belonging to older time intervals should also have older radiometric dates.

Term	Example	Typical Time Range in Years
Eon	Phanerozoic	Hundreds of millions to billions
Era	Paleozoic	Hundreds of millions
Period	Permian	Tens of millions
Epoch	Early Permian	Millions
Age	Leonardian	Hundreds of thousands to millions

Figure 8-7. Terms in the hierarchical time scale used to express geologic time, along with the typical time range associated with each term.

This test, conducted over and over with hundreds of thousands of samples, has repeatedly confirmed the accuracy of the *order* shown in the geologic column and the accuracy of radiometric dating.

Example of Dating a Rock Layer in the Grand Canyon

In 1910, a United States Geological Survey geologist named Nelson Darton described the Kaibab Formation of the Colorado Plateau region and determined, from what was then known about fossils, that it was deposited in the latter part of the Paleozoic Era. Over the ensuing three decades, subsequent geologic investigations of the Kaibab Formation found particular species of fossil worms, sponges, crinoids, echinoids, bryozoans, brachiopods, pelecypods, scaphopods, gastropods, cephalopods, crustaceans, and fish remains that narrowed the time more specifically to the Permian Period (still within the late Paleozoic). Fossils beneath the Kaibab Formation are distinct from those in the Kaibab, and fit well into earlier periods.

Permian rocks are found on all the continents of the world, having first been described in the Perm region of Russia in 1841 by the pioneering British geologist Sir Roderick Impey Murchison. Murchison in 1841 and Darton in 1910 did not know the absolute time frame of the Permian Period. But in 1911, a young British geologist named Arthur Holmes published the first absolute date for a rock, based upon the uranium-lead radiometric method. By the 1950s, radiometric dating was being applied to determine absolute dates for the geologic time scale, including igneous rocks in layers identified as Permian, from multiple locations around the world. The Permian Period is currently identified as beginning 299 million years ago and ending 252 million years ago. This process of combining relative dating with absolute dating, repeated successfully all over the world, is the basis for ages now assigned to layers within the Grand Canyon.

Relative Dating and Flood Geology

Given the scale of disagreement between conventional geology and flood geology, one might reasonably think that flood geologists reject the principles of relative dating and the geologic column. There is basic agreement, however, on the order of events. Recall the discussion back in Chapter 3, where terms for geologic time periods are accepted by flood geologists but are made to fit around and within a one-year flood model. Flood geologists routinely make use of relative dating principles, though they often leave out essential details for a full understanding of the sequence of events. Even more surprising, in spite of their many accusations that the geologic column does not exist entirely in one place (early geologists never claimed it did) or that it is based on evolutionary assumptions (it predates Darwin), is the fact that most flood geologists accept the basic construction of the geologic column. Consider this quote from Henry Morris in the seminal work *Scientific Creationism*: "Creationists do not question the general validity of the geologic column, however, at least as an indicator of the *usual* order of deposition of the fossils...." General agreement is also evident from the many ways in which flood geologists attempt to explain (not disagree with) the observed order of layers and fossils with flood mechanisms. These explanations will be revisited in later chapters.

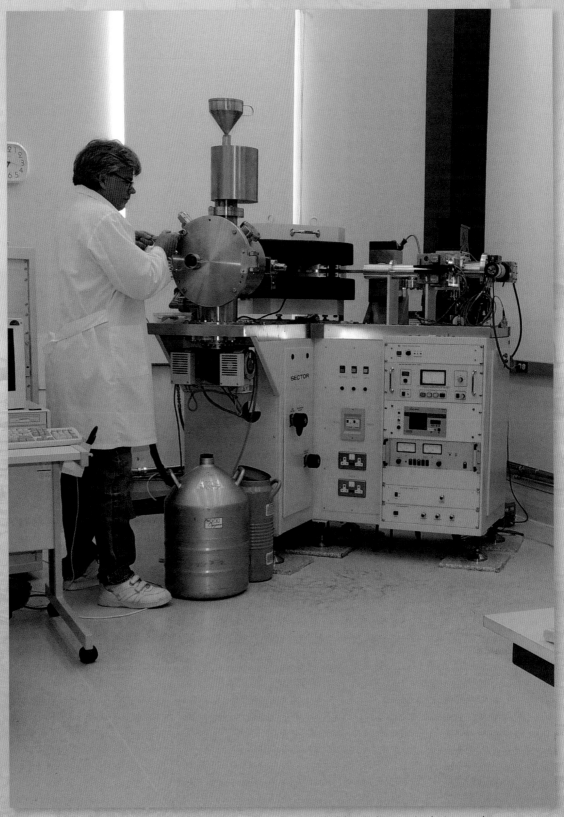

Thermal ionization mass spectrometer (TIMS) used for radiometric dating at the Radiogenic Isotope Lab at the University of New Mexico. *Photo by Victor Polyak.*

CHAPTER 9

SO JUST HOW OLD IS THAT ROCK?

by **Roger Wiens**

How Does Radiometric Age Dating Work?

HOW OLD ARE THE ROCKS OF THE Grand Canyon? Prior to the 1900s, the only tools available for determining the age of a rock were the relative-dating methods described in the previous chapter. One could confidently say that the Kaibab was younger than the Coconino (Principle of Superposition), but determining the actual age of each layer was out of reach. Starting in the early 1900s, however, a number of methods were developed to determine the ages of igneous rocks, independently of their surroundings, by "counting" the atoms produced by radioactive decay. These are called *radiometric dating* methods.

Most of us are familiar with *elements*. Each atom has chemical and atomic properties

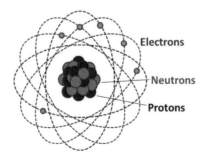

Figure 9-1. Classical depiction of the atom. Isotopes of a given element differ in the number of neutrons (neutral particles) in the nucleus, but always have the same numbers of protons.

that classify it as belonging to a certain element, like carbon, silicon, oxygen, or iron. These properties result from the number of electrons around the atom's nucleus, which matches the number of protons in the nucleus. All atoms of the same element have the same number of protons. However, atoms of the same element can have different numbers of neutrons. Atoms with the same number of protons but variable numbers of neutrons are called *isotopes* (Fig 9-1). For example, carbon always has 6 protons, but the number of neutrons can vary. Carbon can have 6, 7, or 8 neutrons, referred to as carbon-12, carbon-13, or carbon-14 (the atomic weight is the sum of the atom's protons and neutrons: here, 6+6, 6+7, and 6+8, respectively). A different number of neutrons does not change a substance's chemical properties, but it does change its *nuclear* properties, such as the stability of the atom against radioactive decay.

Figure 9-2. Hourglass. *Photo by S. Sepp.*

| Initial State | After 1 Half-Life | After 2 Half-Lives | After 3 Half-Lives | After 4 Half-Lives |

Figure 9-3. The rate of radioactive decay follows the "half" rule. In each interval, called a half-life, half of the remaining "parent" isotopes (purple), on average, decay to become "daughter" isotopes (white).

So in talking about radiometric dating we need to refer to isotopes, not just elements.

The isotopes present in nature are usually stable and do not change. However, some isotopes are not completely stable. Radioactive atoms change, or "decay," from their initial state — which we call the *parent isotope* — to become another isotope, called the *daughter isotope.* Some of these daughter isotopes undergo further decay to become yet another daughter isotope. Radioactive atoms decay at a predictable rate. In a nutshell, the passage of time can be measured by the reduction in the number of parent isotopes and the increase in the number of daughter isotopes.

One feature that makes radiometric dating different from the timepieces we are used to is the concept of *half-lives.* In an hourglass, the rate of sand falling through the neck of the device is constant right up until the very end (Fig 9-2). With an old-fashioned clock, the hand moves just as fast at the top of the hour as at half past the hour. Radioactive decay is different; the number of decaying atoms declines as fewer atoms are left to decay. To measure time with these systems, we define a half-life as the time it takes for half of the atoms to decay. If it takes a certain length of time for half of the original atoms to decay, it will take the same amount of time for half of the remaining atoms to decay. In the next interval, with only a fourth remaining, only one-eighth of the original total will decay (Fig 9-3). By the time ten of these intervals, or half-lives, have passed, less than one-thousandth of the original number of radioactive atoms is left. With the proper conversion from half-lives to actual time (e.g., in years), radioactive decay yields an accurate time measurement.

The simplest view of radiometric dating is that the loss of parent isotopes and gain of daughter isotopes

indicates the age of the object. For an igneous rock, the "age" means the time since the rock crystallized from a magma, either inside the Earth or as part of a volcanic system. Young Earth advocates frequently claim that radiometric dating doesn't work because there are usually some daughter atoms already present at the time that the magma crystallizes — akin to having some sand grains already in the bottom of the hourglass when it is flipped over. Scientists are well aware of this, however, and have discovered accurate ways to determine how many of the daughter atoms existed when the clock was started. With this knowledge, scientists can determine rock ages properly.

Old rocks (which turn out to be most of them) are dated by radiometric methods that use long half-lives of millions of years. These long half-lives are similar to the ages of the rocks themselves, and this is necessary so that the parent isotope does not completely run out. The elements involved include uranium, lead, rubidium, strontium, neodymium, samarium, potassium, and argon, to name a few.

Let's summarize what radiometric dating has revealed about the ages of rocks on Earth and elsewhere. From our current understanding of how our solar system was formed, the oldest rocks are meteorites that strayed from the asteroid belt into Earth's gravitational field. Meteorites commonly date to about 4.5 billion years old. Rocks from the Moon range from nearly 4.5 billion years old to just over 3 billion years old for its younger lava flows. Martian meteorites also date to over 4 billion years old in some cases, but are less than 100 million years old in other cases, indicating that the Red Planet may be slightly active volcanically even today.

No rocks on Earth are as old as the meteorites, in part because the Earth is a very dynamic planet

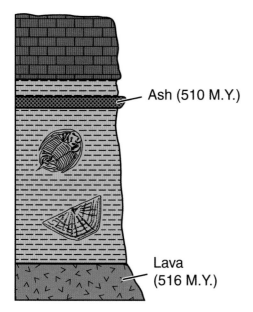

Ash (510 M.Y.)

Lava
(516 M.Y.)

Figure 9-4. If the fossils shown were deposited on top of a 516-million-year-old lava flow and sit beneath a 510-million-year-old ash deposit, the age of the fossils can be established as being somewhere between 510 and 516 million years.

Figure 9-5. Lava flows in the western Grand Canyon. These occurred after the canyon was formed and have been dated to an age within the last several million years. *Photo by Tim Helble.*

with active processes creating and breaking down rocks. Rocks from the ocean floor are generally less than 200 million years old, but much older rocks are found in places on the continents. The oldest are found in continental "shields," in parts of Canada and Australia. Many of those rocks are well over 3 billion years old, and some have been dated to over 4 billion years since their crystallization from a magma.

Sedimentary rocks usually cannot be directly dated by radiometric methods, because the grains that make up the rock do not reflect the time when the grains (such as sand or mud particles) were cemented together as a rock. So how do we get ages for fossils

Figure 9-6. Some different types of radiometric and nonradiometric dating techniques and ages covered by each technique. Nonradiometric and radiometric techniques give compatible dates, confirming the reliability of radiometric dating.

Dating Techniques	Ages Covered by Technique
Radiometric Techniques:	
Uranium-lead	4.5 billion years to 1 million years
Rubidium-strontium	4.5 billion years to 60 million years
Potassium-argon	3 billion years to 3 million years
Argon-argon	3 billion years to about 2,000 years
Uranium series	400,000 years to a few centuries
Carbon-14	40,000 years to about 50 years
Non-Radiometric Techniques:	
Electron spin resonance	1 million years to 1,000 years
Thermoluminescence	500,000 years to a few years
Amino acid racemization	300,000 years to 500 years
Tree ring counting	12,000 years to 1 year

Figure 9-7. The Dead Sea Scrolls contain portions of the Hebrew scriptures, such as this section from the Psalms, radiocarbon dated to between 350 BC and 100 AD. *Photo by Shai Halevi, Full Spectrum Color Image, Courtesy of the Israel Antiquities Authority.*

in sedimentary rocks? The answer is to take advantage of nearby igneous rocks that can be dated (Fig 9-4). As an example, if lava or volcanic ash layers are found above and below a fossil-bearing sedimentary rock, the volcanic layers can be dated, which provides an upper and lower age boundary that the sedimentary layer must fit within. In the Grand Canyon, the radiometric age for a volcanic ash bed at the base of the Bass Formation (at the bottom of the Unkar Group), was found to be 1,255 ± 2 million years and the age of the Cardenas Lava at the top of the Unkar Group has been determined to be 1,104 ± 2 million years (U-Pb method). This means that the Bass Formation, Hakatai Shale, Shinumo Quartzite, and Dox Formation were deposited over a period of about 150 million years.

Radiometric dating can be applied to most igneous rocks, such as the granite dikes or ancient lava flows in the Grand Canyon (Fig 9-5). However, partial resetting of a radiometric clock can occur if the rock is heated enough to allow migration of the parent and/or daughter isotopes. Such heating over long periods of time underground is called *metamorphism* (literally "reshaping"), and rocks that have clearly been altered are called *metamorphic*. Much of the crystalline basement of the Grand Canyon is schist, a type of metamorphic rock derived from sedimentary rock. These rocks are generally not useful for dating, unless what you are interested in is the timing of metamorphic resetting.

In normal life, if you really want to be sure of the time, you check more than one clock. Because different isotopes respond differently to metamorphic heating, one of the best ways to ensure an accurate date is to check for agreement among two or more dating methods (Fig 9-6). In most studies, various methods

yield consistent ages, increasing our confidence in the validity of the calculated ages. For example, in 1986, a study was completed that compared the record of changes in iron mineral alignment in Unkar Group sedimentary layers to the changes in alignment of iron minerals in datable igneous rocks outside the Grand Canyon area. These changes record reversals in the Earth's magnetic field. This study found that the Unkar Group layers started accumulating 1,250 million years ago. This is essentially the same as the radiometric age found for the ash layer at the bottom of the Unkar Group in 2005.

Radiocarbon (C-14) Dating

Radiocarbon, or carbon-14, dating uses an isotope of carbon that is constantly produced in the Earth's upper atmosphere. Radiocarbon is used to date dead organisms that at one time breathed (or in the case of plants, respired) the air and incorporated the carbon-14 isotope into their tissue. Carbon-14 in living organisms continually decays and is continuously replenished, but once a plant or animal dies, no more carbon-14 is added, and the carbon-14 that is in the tissue slowly decays with a 5,730-year half-life. For radiocarbon dating, one only measures the parent carbon-14 abundance, not the daughter isotope. Samples that can be dated with radiocarbon include materials like wood, papyrus, natural fabric, and animal bones and tissues. The Dead Sea Scrolls are an example of an archeological artifact dated by the radiocarbon method (Fig 9-7).

A serious misconception is that rocks are dated by radiocarbon. With the exception of very young limestone deposits, radiocarbon is not a useful way to date rocks. The radiocarbon half-life is much shorter than the age of most rocks, which means that virtually all their radiocarbon has already decayed away. After about forty thousand years (about seven half-lives), less than 1 percent of the original radiocarbon is left. Radiocarbon is also a special case that is more sensitive to contamination than the radiometric methods used for rocks. The reason is that the air, our skin, hair, pollen, and other materials around us contain much more carbon-14 than the amount being measured in very old samples. As smaller and smaller amounts of the original carbon-14 remain in an ancient artifact, the potential for errors due to contamination increases substantially, so that large errors can result when attempting to date the oldest samples. This is not much of a problem for tissue or plant material produced within the last 20,000 years (lots of original carbon-14 still left), but the potential for error grows significantly with older artifacts.

Some Young Earth advocates insist on trying to use radiocarbon to date old rocks, coal, or diamonds, with meaningless results. During sample processing, small amounts of radiocarbon are inevitably incorporated from the air in the lab, so that a truly zero reading

Figure 9-9. Ash-flow deposits destroyed the city of Pompeii in 79 AD. The bodies were not preserved, but were found as voids in the solidified ash. The voids were filled with plaster and excavated in the nineteenth century. *Photo by Lancevortex.*

is never obtained. This is neither puzzling nor disconcerting to scientists. The assertion by flood geologists that "measurable radiocarbon" in coal and diamonds is evidence for a young Earth is simply misleading.

How Reliable Is Radiometric Dating?

Radioactive materials have many different uses, notably in medicine and in generating electrical power. Half-lives have remained constant and have proven to be trustworthy in these industries over a long period of time. The Mars rover that I am working on – now trekking across the Martian landscape – is powered by an isotope of plutonium that has a half-life of 88 years (Fig 9-8). If we are fortunate enough to still be operating the rover that long from now, we can be assured that it will have half the heat energy it has now. Familiarity with radiometric half-lives in different industries gives us confidence in using them to determine the ages of rocks.

There is also agreement between the dates recorded for ancient historical events and the radiometric ages measured for rocks or other articles from those events. Probably the best-known example is the eruption of Mount Vesuvius in Italy nearly 2,000 years ago, which destroyed the cities of Pompeii and Herculaneum in 79 AD (Fig 9-9). Rocks produced by this eruption have been correctly dated using the argon-argon method to within a few years of the event. There is no evidence of any half-lives changing over time.

Young Earth advocates will often concede the validity of radiometric dating for samples only a few thousand years old, but insist that they are not reliable

Figure 9-8. NASA's one-ton Mars rover, Curiosity, is powered by a radioactive isotope with an 88-year half-life. *Photo by NASA/JPL-Caltech/Malin Space Science Systems.*

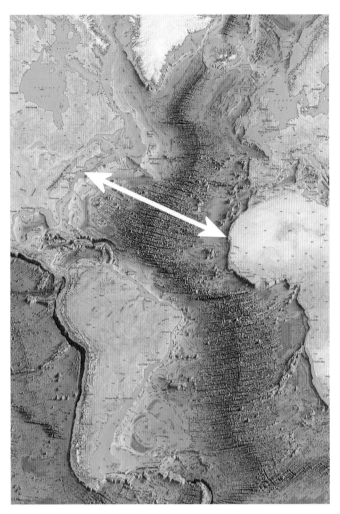

Figure 9-10. The Mid-Atlantic Ridge. The arrow represents a separation of 3,480 miles along the direction of movement. *Reproduced by permission of Marie Tharp Maps, LLC, 8 Edward Street, Sparkill, New York 10976.*

plates making up the Earth's outer crust. If we consult a map of the Atlantic Ocean floor (see Fig 9-10, to the left), a large mountain ridge – aptly named the *Mid-Atlantic Ridge* – is readily visible, exactly dividing North and South America from Europe and Africa. Lava welling up along the ridge is constantly adding new crust to the ridge and attests to continuing modern-day separation of the continents and expansion of the ocean floor. The ages of rocks from the ocean crust, determined using radioactive methods, are consistent with this observation; the farther one moves away from the ridge, the older the rock.

Maximum ages of about 180 million years for the Atlantic Ocean crust are obtained at the edges of the continents. If we measure the distance between continents, we can use the distance and the maximum age to come up with an average rate of separation. Doing this along the arrow shown on the map (3,480 miles over 180 million years) yields an average rate of 1.2 inches per year. If this exercise is repeated at intermediate locations along the same arrow, the intermediate ages and distances yield spreading rates that appear to have varied only slightly over time, with rates ranging from about 1.1 to 1.7 inches per year.

Now for the test. Satellite stations on different continents allow precise measurement of distances, down to the scale of inches or less. Long-term measurements of the relative positions of North America and North Africa document a current spreading rate of approximately 1 inch per year, a value in remarkable agreement with the radiometrically determined rates.

Consider the implications here. Flood geologists claim that the North American and African plates began moving violently and rapidly apart as subterranean water burst to the surface through fissures at the onset of Noah's flood. Plate movement then gradually slowed down over the next several thousand years to their current measured rates. For this to be true, two entirely unrelated processes – the decay rate of radioactive isotopes and the rate of plate spreading – must have slowed down at *exactly the same rate* in order to only appear to verify the accuracy of radioactive dating methods!

Thousands of rock samples are dated each year, and in the vast majority of these cases there is agreement among different dating methods, and among

for older rocks because the methods are based on untestable and often questionable assumptions. Dates are only reliable *if* radioactive decay rates have not changed over time, *if* the systems remain closed, *if* the rocks were not heated too much, and *if* samples have not been contaminated. With so many *if's*, how can we ever be confident that radiometric ages are accurate? What if multiple radiometric techniques agree only because the rate of decay of the different isotopes has been slowing, by some unidentified process, at the same proportional rate? If we genuinely could not test our assumptions, our confidence in the reliability of the results might truly be suspect. But the claim that these assumptions are not testable is utterly false.

A simple method for testing assumptions and verifying the accuracy of radioactive dating makes use of our knowledge of plate tectonics – the movement of

Figure 9-11. Mount St. Helens building up a central cinder cone after its 1980 main eruption. The dark surfaces in the foreground are ash flows from that eruption. *Photo by U.S. Geological Survey.*

rocks in the same area, presenting a clear picture of an old Earth.

Although ample tests have been done to confirm the accuracy of radiometric dating, some flood geologists still claim that radiometric dating gives inaccurate or irreproducible results. A popular example of this claim involves Mount St. Helens in Washington, which erupted in 1980, sending volcanic ash flows racing down the mountainside. In 1986, a fresh cone of magma growing in the crater was sampled by Young Earth advocates and submitted for dating (Fig 9-11). A potassium-argon date of 300,000 years was calculated for this sample, which is now repeatedly featured in flood geologist lectures, videos, field trips, web sites, and museum exhibits as evidence against the reliability of radioactive dating. However, two critical pieces of information are *left out*. First, it is common for fragments of old rock from the subsurface to become entrained in the upwelling lava. Indiscriminate sampling of the lava could very well result in a mixture of new and old rock, yielding an age much older than that of the lava flow. The second piece of information left out is that the dating technique used (potassium-argon) has long been recognized by geologists to yield inaccurate results for recent lava flows – not because

"they don't give the right answer," but because of known processes at work in this environment. Methods such as argon-argon dating take these processes into account, and do in fact yield reliable ages (such as accurately dating the Mount Vesuvius eruption mentioned above). Radiometric dating techniques *are* accurate. This has been proven over and over again.

Back in the Canyon

Most of the layers exposed in the Grand Canyon are sedimentary rocks. Dating these requires a combination of radiometric and relative dating methods, as previously described. Where the sedimentary rocks are associated with igneous rock, dates determined for the igneous rocks can be used to constrain the possible ages of the sedimentary rocks. Some of the youngest rocks in the canyon are composed of lava that flowed over the rim and down to the river, forming lava dams. Eventually, the backed-up Colorado River breached the dams and eroded or partly eroded them away, leaving only telltale lava remnants along the sides of the canyon (see Fig 9-5, page 91). These lava dams are much younger than the underlying rock layers and

Method	Age
Potassium-argon	840±164 million years
Rubidium-strontium	1,055±46 million years
Uranium-lead	1,249±130 million years
Samarium-neodymium	1,375±140 million years

Figure 9-12. Different dates for the igneous intrusion within the Hakatai Shale, shown with the uncertainties.

help us understand the age of the canyon itself, but they tell us little about the age of the rock layers into which it was carved. Most of the lava dams have formed only within the last few million years.

There are also very ancient igneous rocks in the lowest sections of the eastern Grand Canyon. Some have been heavily metamorphosed, so they are not readily dated. However, one rock formation from the Grand Canyon Supergroup can be dated: the Cardenas Lava. Various radiometric dates on this lava give it an age of about 1 billion years.

Although we have already demonstrated conclusively that radiometric dating works, it is worth revisiting the subject for a moment to address the commonly quoted "inconsistent" (or "discordant") dates reported by flood geologists for igneous rocks in the Grand Canyon. Two separate Young Earth investigations, one from an igneous intrusion within the Hakatai Shale, and one from the much younger lavas that flowed over the canyon rim, illustrate ways in which sampling, dating, and reporting are handled to create the appearance of unreliability.

Dating an Igneous Intrusion:
Leaving Out Information

Flood geologists collected samples from an igneous intrusion within the Hakatai Shale of the Grand Canyon Supergroup and submitted them for dating using four different radioactive dating methods, with results that appear to be inconsistent (Fig 9-12).

Young Earth references to this work typically fail to include two important pieces of information. First, geologists have known for a long time that argon, a very mobile gas, leaks from reheated

rocks much more readily than strontium, lead, or neodymium do. Loss of argon (the daughter isotope) means the sample will look younger than it truly is (note the rock does not look falsely old, it looks falsely *young*). A younger calculated age for the potassium-argon method is thus evidence of heating and argon leakage, and therefore is not the most reliable method for dating this intrusion. The second piece of information that is often left out involves the uncertainties (the ±

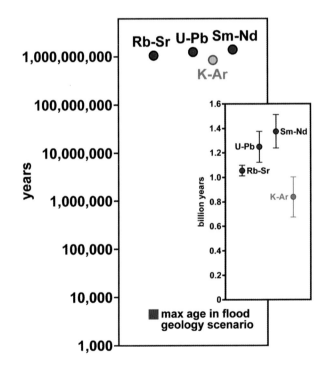

Figure 9-13. Reported dates by flood geologists for samples from the igneous intrusion within the Hakatai Shale. The vertical lines represent the measurement uncertainty (Rb-Sr: rubidium-strontium; U-Pb: uranium-lead; Sm-Nd: samarium-neodymium; K-Ar: potassium-argon). Grayed-out K-Ar point indicates unreliable date due to argon leakage. The blue square represents the maximum age for the igneous intrusion suggested by the Young Earth literature.

values that define the prediction range). Though uncertainties are normally reported in scientific communications, they are often left out of Young Earth literature to make the dates seem more discordant than they actually are. Consider the plot on the right of Fig 9-13, showing bars to indicate the prediction ranges. In this figure, we see that the U-Pb and Sm-Nd methods actually agree within the prediction ranges, and uncertainty for the Rb-Sr result places it within about 10 percent of the uncertainty range of the Sm-Nd value. These are hardly the wildly divergent results claimed by flood geologists.

Western Grand Canyon Lavas:
Improper Sampling

The much more recent lavas that poured over the rim of the canyon are found today in the western Grand Canyon and are radiometrically dated to a few million years in age. Attempts to date some of these flows have also been made by flood geologists. In one widely cited study, a particular approach was employed, called an *isochron method*, that requires collecting multiple samples from the *same* rock unit.

In the flood geology study, samples were instead collected from *four different flows*, guaranteeing meaningless "ages."

Conclusion

To summarize this chapter, radiometric dating is based upon physical principles that we can trust — most importantly, on the predictability of radioactive decay. We use this predictability to construct nuclear reactors, improve medical devices, and even power a rover on a planet millions of miles away. Radiometric dating has proven to be accurate in dating ancient historical events such as the eruption of Mount Vesuvius. The same methods show that the Grand Canyon cannot be younger than a million years old, based on the age of lava that flowed into the gorge, and that some of the basement rocks into which the canyon was carved are more than one billion years old.

COULD DECAY RATES HAVE BEEN MUCH FASTER IN THE PAST?

Increases in radioactive decay rates in the past — more than a million times the currently observed rates — have been suggested by some flood geologists as a way to reconcile the billion-year radiometric ages with their idea of a six-thousand-year-old Earth. But this ignores a fundamental property of radioactive decay: it produces heat. The slow radioactive decay of materials in the Earth keeps the Earth's interior hot, powering active volcanoes. Increase that heat by a factor of a million or more, and the Earth's surface would quickly become completely molten (Fig 9-14). There is no reasonable way to deal with the tremendous heat generation of accelerated decay, except to acknowledge that it didn't happen that way.

Figure 9-14. A picture of a molten planet based on an artist's concept from NASA/JPL-Caltech.

The unconformity between the light colored Coconino Sandstone and the reddish Hermit Formation.
Photo by Bronze Black.

CHAPTER 10

MISSING TIME
GAPS IN THE ROCK RECORD

by **Stephen Moshier** and **Carol Hill**

GEOLOGISTS HAVE OFTEN COMPARED geologic layers to pages in a book: each layer tells a story that is continued in the next, overlying layer. However, time gaps in the geologic story are like pages or whole chapters ripped out of the binding of the book. A time gap between adjacent rock units is represented by an *unconformity:* the contact between two layers of rock where material was eroded away or where time passed without additional deposition of sediment. Usually an unconformity implies erosion, but not always.

In previous chapters we have described the most obvious unconformities in the Grand Canyon, such as the Great Unconformity (Chapter 4), and the lower unconformity between the Vishnu Schist and the overlying Supergroup. But we have not described how we recognize these features or determine how much time is missing. We'll address these issues by taking a closer look at five of the 19 identified unconformities in the Grand Canyon (all 19 are shown in Fig 10-1 on following page).

How Much Time?

Estimating the length of time that is missing from the rock record is more difficult than simply identifying that a gap in time is present. Where igneous rock is eroded, radiometric dating can at least yield the age of the underlying rock formation, though not the timing of the later erosion. In the case of sedimentary rock in places like the Grand Canyon, estimates are based largely on a comparison of the local fossil record with the global record. Fossil assemblages identifying particular geologic time periods above and below an unconformity, combined with knowledge of the ages of those time periods (from radiometric dating methods), allow an estimate of how much time is missing.

HOW CAN TIME SIMPLY BE "MISSING?"

Flood geologists often cast doubt on the validity of unconformities by implying that geologists think nothing was happening during these vast stretches of time, with exposed rocks just sitting idly for millions of years – which is clearly implausible. But this is far from true. In many places, it is likely that thick sequences of deposits formed over these time periods, but were later eroded away. The missing time is not typically a record of inactivity; it is a record of *removed* history.

Top of Kaibab Formation	Beyond canyon rim, Permian Kaibab Formation locally covered by Triassic Moenkopi Formation. Estimated 15 million years (my) gap.
Hermit Formation to Coconino Sandstone	Deep (up to 20 ft) cracks in Hermit filled with sand. Much rock is missing at top of Hermit in Grand Canyon compared to other Colorado Plateau localities. Estimated hundreds of thousands of years gap.
Esplanade Sandstone to Hermit Formation	Local erosion surface at top of Esplanade; 30-50 feet of relief. Estimated few my gap.
Wescogame Formation to Esplanade Sandstone	Erosion surface at top of Wescogame; local relief up to 50 feet. Estimated few my gap.
Manakacha Formation to Wescogame Formation	Scoured top of Manakacha, with conglomerate clasts reworked into base of Wescogame. Minor relief evident. Estimated 10 my gap.
Watahomigi Formation to Manakacha Formation	Estimated 2.5 my gap.
Surprise Cyn Formation to Watahomigi Formation	Localized low-angle unconformity with conglomerate at base. Estimated 10-15 my gap.
Redwall Limestone to Watahomigi Formation	Estimated 15 my gap.
Redwall Limestone to Surprise Cyn Formation	Network of channels cut into top of Redwall filled with Surprise Cyn Fm. Channels up to ½ mile wide by 400 feet deep. Estimated few my gap.
Temple Butte Formation to Redwall Limestone	Erosion surface at top of Temple Butte, with conglomerate containing reworked Temple Butte at base of Redwall. Estimated 30 my gap.
Redwall Limestone to Muav Limestone	Estimated 160 my gap.
Muav Limestone to Temple Butte Formation	Upper surface of Muav cut by channels (up to 100 feet deep); filled with Temple Butte. Estimated 130 my gap.
The "Great Unconformity" Cambrian Tapeats Sandstone over Precambrian rocks.	Conglomerate at base of Tapeats over top of Precambrian rocks. One billion year gap of Tapeats over crystalline basement in western Grand Canyon; estimated 220 my gap between Chuar Group and Tapeats in eastern Grand Canyon.
Kwagunt Formation to Sixty Mile Formation	Conglomerate in Sixtymile over erosion surface on top of Kwagunt. Estimated gap up to 50 my.
Nankoweap Formation to Galeros Formation	Erosion surface at top of Nankoweap. Estimated few millions of years gap.
Unconformity between lower and upper members of the Nankoweap Formation.	Missing time unknown.
Cardenas Lava to Nankoweap Formation	Nankoweap over slightly tilted and eroded Cardenas Lava. Estimated gap up to 300 my.
Hakatai Formation to Shinumo Quartzite	Truncated cross-beds and channel deposits in Hakatai. Missing time unknown, as much as 50 my.
Early Proterozoic Crystalline Basement to Bass Limestone	Estimated 410 my gap in time.

Figure 10-1. Nineteen unconformities exposed in the Grand Canyon (my=million years).

Figure 10-2. The Great Unconformity in the eastern Grand Canyon, showing the Tapeats Sandstone and Bright Angel Shale overlying inclined red Supergroup rocks (Chuar strata). The high point (arrow) is an ancient butte composed of resistant Shinumo Quartzite that stood above the flat-lying plain before the advance of the Tapeats sea over the land. *Photo by Eben Rose.*

The total missing time for just the unconformities in layers above the Great Unconformity adds up to about 190 million years. This means that out of the total time represented by the Paleozoic rocks of the Grand Canyon – from the 525-million-year-old Tapeats Sandstone up to the 270-million-year-old Kaibab Formation (a period of about 255 million years) – roughly 75 percent of the rock record is missing. This is why flood geologists consider it so important to deny that any unconformities exist in rock above the Great Unconformity – because unconformities would mean repeated and extended periods of exposure and erosion in the middle of their continuous year-long flood.

Recognizing Unconformities

There are a number of ways to recognize unconformities. Five characteristic examples are provided from the Grand Canyon: (1) Tilted and Truncated Layers, (2) Weathered Rock in Overlying Unit, (3) Missing Fossils, (4) Filled-In Channels, and (5) Filled-In Caves and Sinkholes (Paleokarst).

1. Tilted and Truncated Layers

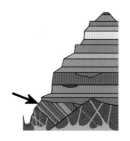

The most obvious evidence of erosion is found when tilted layers abruptly terminate against overlying horizontal layers, as is vividly illustrated by the Great Unconformity in the eastern Grand Canyon where the Tapeats Sandstone overlies tilted Supergroup rock (Fig 10-2). The only way to generate such a contact is to tilt originally horizontal layers, erode material off the top to a relatively flat surface, and then deposit new material horizontally on top. The Great Unconformity is one of the few features that flood geologists acknowledge as a genuine unconformity, and they put it at the beginning of the flood when the land supposedly was being scoured by floodwaters. Remember, to make this unconformity, over 12,000 feet of Supergroup sedimentary rocks were (1) deposited, (2) buried, (3) slightly metamorphosed, (4) faulted and tilted, (5) uplifted and subjected to erosion, and finally (6) covered by the younger sequence of Cambrian-age strata. Flood geologists insist that all of this happened in some 1,650 years, between the creation week and the start of the flood!

2. Weathered Rock From Lower Unit Contained in Overlying Unit

If we move westward along the Great Unconformity to where the Tapeats Sandstone overlies crystalline rock, we can find "rip-up clasts" at the contact between these two units (Fig 10-3). We find similar clasts at the base of the Surprise Canyon Formation layers. The rock fragments in the Surprise Canyon Formation are clearly from the underlying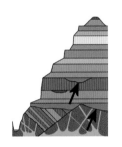

Redwall Limestone – they match it, right down to the distinctive fossils found elsewhere in the Redwall. How did these rock fragments get there? The presence of Redwall fragments in the overlying rock makes perfect sense if the top of the Redwall was eroded, leaving weathered pieces on the surface or in collapsed caves during subsequent deposition of the Surprise Canyon sediments. If deposition was continuous (i.e., if there was no unconformity), how did chunks of hard Redwall Limestone form rapidly and find their way up into the Surprise Canyon Formation without defying gravity? (Hint: They couldn't).

3. Missing Fossils

We have already noted that long before Darwin, geologists observed a regular sequence of fossil organisms in the rock record (remember our Principle of

Figure 10-3. Great Unconformity contact of Vishnu Schist with the overlying Tapeats Sandstone in Blacktail Canyon. This unconformity represents a time gap of 1.2 billion years. Inset: the white quartz clasts along the contact were once present as veins within the schist. *Photos by Wayne Ranney.*

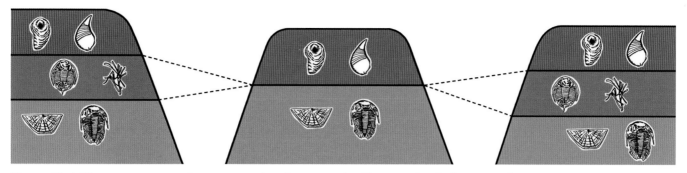

Figure 10-4. The center outcrop has an unconformity, recognized because fossils that normally are separated in time (as seen in the left and right outcrops) occur here in two adjacent layers. Note: This illustration shows index fossils that are not specific to the Grand Canyon.

Faunal Succession). In some locations, layers are found with suites of fossils in the expected order but with one or more entire fossil groups missing (Fig 10-4). The abrupt transition from an older fossil assemblage to a much younger one indicates a period when either no sediment was deposited or material was eroded away. In the case of the Muav-Temple Butte contact, there are only Cambrian-age invertebrate fossils in the Muav Limestone, whereas Devonian-age fish and invertebrate fossils are present in the Temple Butte Formation. At this location, the time gap between the Muav Limestone (~505 million years) and the Temple Butte Formation (~375 million years) represents about 130-135 million years of missing history.

4. Filled-In Channels

An obvious clue to the existence of an unconformity is the presence of channels carved out of a lower stratigraphic unit that are filled either with material from an upper unit or with unique deposits. Good examples are found at the tops of  the Muav and Redwall formations. Channels carved in the surface of the Muav are filled in with Temple Butte deposits that are distinct from the overlying Redwall Limestone (Fig 10-5). Channels carved in the top of the Redwall are filled in with deposits from the Surprise Canyon Formation that are distinct from the overlying Supai Group (Fig 10-6). Flood geologists often point out that the Grand Canyon's layers appear to be pancake flat, but these Surprise Canyon channels can be up to 400 feet deep!

Some of these channels contain clear evidence of having once been above sea level. The deep channels in the Redwall are now filled in with sediment containing land fossils at the bottom and marine fossils on top. This is possible only if the Redwall Limestone was above sea level for an extended time and was later submerged.

The bottom layers of the Surprise Canyon Formation (the rock filling the channels carved into the Redwall) are river deposits that contain fossil plant material, including giant seed-fern *Lepidodendron* trees (Fig 10-7). These trees grew in low-lying swamplands about 320 million years ago, and reached heights of 90 feet and had trunks as thick as 6 feet in diameter. These fossils unequivocally represent a land environment, not a marine environment. Rivers were flowing over the landscape at this time.

The middle and upper layers of the Surprise Canyon Formation are limestones containing marine fossils – corals, brachiopods, echinoderms (e.g., organisms like starfish), and sharks' teeth. These fossils attest to a shallow sea advancing over the land and filling in the valleys, sinkholes, and caves with marine limestone.

5. Filled-In Caves and Sinkholes (Paleokarst)

Karst, a term applied to a landscape altered by caves and collapse features (sinkholes), is found in some Grand Canyon layers, particularly in the Redwall

Figure 10-5. A channel in the Muav Limestone, filled in with material from the Temple Butte Formation. *Photo by Robert Leighty.*

Figure 10-6. A 400-foot-deep channel in the Redwall Limestone, Quartermaster Canyon, western Grand Canyon, filled in with Surprise Canyon Formation rock and overlain by rock from part of the Supai Group. *Photo by George Billingsley.*

Limestone. We mentioned above that river channels are carved into the surface of the Redwall Limestone, but there is more to this story. The upper part of the Redwall Limestone also exhibits a *paleokarst* terrain (paleo = ancient), where sinkholes developed on the surface of the Redwall Limestone as caves formed and collapsed. Later deposition filled in many of those caves and sinkholes with new sediments, today identified as the Surprise Canyon Formation – the same material that fills in the channels in the Redwall Limestone.

Figure 10-7. *above left: Lepidodendron* tree. *above right: Lepidodendron* tree trunk fossil, Surprise Canyon Formation, Granite Wash Park, western Grand Canyon. *Photo by George Billingsley.*

How do geologists determine what this landscape looked like millions of years ago, since no one was there? How can they claim to know that caves and sinkholes pitted the landscape and that rivers flowed westward to the sea on top of the Redwall Limestone? They know because they can see the remnants of old river networks, sinkholes, and caves preserved in the rock (Fig 10-8). The Surprise Canyon Formation does not lie in a horizontal sheet over the Redwall Limestone. The deep channels filled with river sediments and land fossils testify to the presence of ancient rivers flowing to the west. Widespread caves (later filled in) attest to the fact that streams also flowed below the ground surface, carving out subterranean passages.

At this time, the landscape looked very much like today's Yucatan Peninsula – where "cenotes" (sinkholes) and caves dot a landscape only a few hundred feet above sea level (described in Chapter 7).

These features host the best uranium-ore deposits in North America. The Orphan Mine on the South Rim of the Grand Canyon was mined for copper in the late 1800s and for uranium in the 1900s (Fig 10-9).

Old Caves Exposed by New Caves

Part of the research done by one of us (Carol Hill) on the Grand Canyon involves sampling in modern caves. In some places, modern cave passages have exposed much older caves that were filled in long ago after their roofs collapsed. The 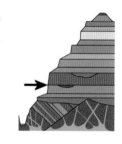 filling material is called breccia. Breccia forms a unique kind of conglomerate, where the individual pieces are jagged rather than rounded. Jagged, angular blocks attest to transport over a short distance or for a brief time before being buried and cemented – consistent with expectations for collapse into an underlying cave. Angular rock fragments in breccias confirm that the layers above the Redwall were already hardened into rock when the caves collapsed – they were not soft sediment from a recent flood (Fig 10-10).

A

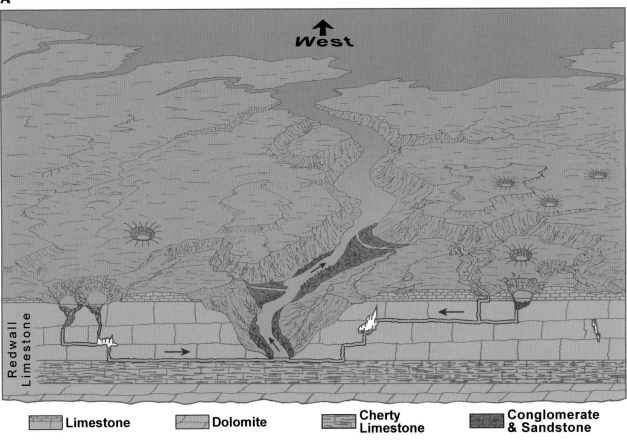

| | Limestone | | Dolomite | | Cherty Limestone | | Conglomerate & Sandstone |

B

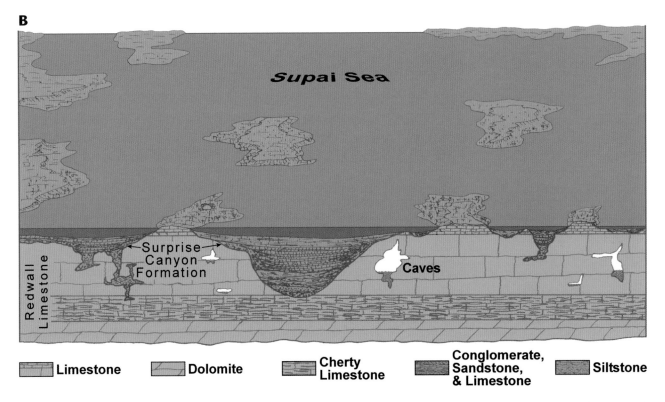

| | Limestone | | Dolomite | | Cherty Limestone | | Conglomerate, Sandstone, & Limestone | | Siltstone |

Figure 10-8. (A) Illustration of what the top of the Redwall Limestone would have looked like when it was above sea level (Late Mississippian Period): a landscape of sinkholes and caves; and (B) later filling of channels with Surprise Canyon deposits as the sea advanced inland again. *Modified from Billingsley and Beus, Geology of the Surprise Canyon Formation of the Grand Canyon, Arizona; used with permission of the Museum of Northern Arizona.*

Figure 10-9. Early Grand Canyon pioneer Ellsworth Kolb negotiating a ladder down the Coconino Sandstone to access the Orphan Mine, circa 1913. This mine, established in a breccia pipe two miles northwest of Grand Canyon Village, was originally a copper claim but later became an important source of uranium during the Cold War. *Photo from Grand Canyon National Park #11344.*

SHOULDN'T EROSION CREATE HILLS AND VALLEYS?

Perhaps the most often cited flood geology "evidence" against unconformities is that erosion should create hills and valleys, while the Grand Canyon contacts are claimed to be horizontal. While some Grand Canyon unconformities are indeed relatively flat, many do exhibit hills and valleys, as shown by the photos in this chapter. We see abundant evidence that these ancient channels and caves were formed by normal surface water processes, not globe-covering flood waters and tsunamis. One of the main premises of flood geology is that the sediment deposited in the flood was still soft when acted upon by tectonic forces (see Chapter 12). But how could caves described in this chapter form in soft sediment and not immediately refill with the surrounding soft sediment slumping into them? Flood geologists point out only the features that appear to support their arguments, not the features that do not.

WHY DOES A PALEOKARST UNCONFORMITY IN THE GRAND CANYON CHALLENGE FLOOD GEOLOGY?

Filled-in caves (paleokarst) require a sequence of events over time. First, limestone was deposited and hardened to rock so that caves could dissolve in the limestone (voids in soft lime mud will not stay open). Next, pieces of the overlying rock fell into the cave voids, eventually filling the caves with rock fragments (breccia). Then the thick layers of the Supai Group were deposited over the Redwall Limestone, compacting and cementing the breccia pieces into rock. Finally, groundwater flow dissolved new caves into the Redwall Limestone to expose the old cave fillings. How could any of these events have occurred within the context of a single-year flood? Such a sequence is not just unlikely during a flood event – it is impossible.

left: **Figure 10-10.** Rock fragments (breccia) exposed in the wall of a modern Grand Canyon cave. *Photo by Bob Buecher.*

An aerial view of the Butte Fault trending to the north, revealing upturned strata to the right of the fault.
Photo by Wayne Ranney.

TECTONICS AND STRUCTURE

The final section of Part 2: How Geology Works is related to the forces that have physically deformed the Grand Canyon's layers, alternately lifting them and causing them to subside, sometimes breaking (faulting) and at other times bending (folding) them. The study of the resulting features is called structural geology. The next chapter, Chapter 11, describes the forces at work at the planetary scale – moving entire continents and ocean basins. Chapter 12 focuses on what can be learned from local-scale faults and folds, with particular attention paid to those features in the Grand Canyon produced from plate movement.

PLATE TECTONICS
OUR RESTLESS EARTH

by **Bryan Tapp** and **Ken Wolgemuth**

Plate Tectonics, the Driving Force

OUR APPRECIATION FOR THE FORCES that cause uplift, subsidence, earthquakes, volcanism, folding, and faulting on our planet has only developed in the last 60 years with the development of our understanding of *plate tectonics*. Plate tectonics is the science that studies the movement of the Earth's crust. Our planet is composed of gigantic plates of rigid crust that are in constant motion (Fig 11-1). Technically, the rigid plate is composed of both crust and the uppermost part of the mantle (jointly referred to as the *lithosphere*), but we'll refer to the crust through this chapter for simplicity.

This movement occurs because the Earth's mantle, below the crust, acts like flexible plastic (think Play-Doh or modeling clay) and is convecting; that is, sections of the mantle are slowly moving in huge circles – rising, drifting laterally, and then sinking (Fig 11-2). While a detailed study of the

history of tectonics at any location is complex, the basic ideas are quite simple.

To understand what the Grand Canyon is telling us about its own tectonic history, we first need to understand the basic driving mechanisms for plate motion and determine whether physical laws give us any "speed limits" on how fast the Earth's plates can move or how fast mountains can rise. Then we can see how these restrictions apply to the claims by flood geologists that the continents spread apart very rapidly during Noah's flood, with mountain ranges rising in just a few months (referred to as "catastrophic plate tectonics" or "runaway subduction").

If we examine the driving forces for plate tectonics, we are left with the simple concepts of gravity, heat, and density. We know that the interior of the Earth is hot and that the temperature of the Earth's surface is relatively cool. This temperature difference, along with gravity, provides the simple mechanics of plate tectonics. The rock material deep inside the Earth is hotter than the material

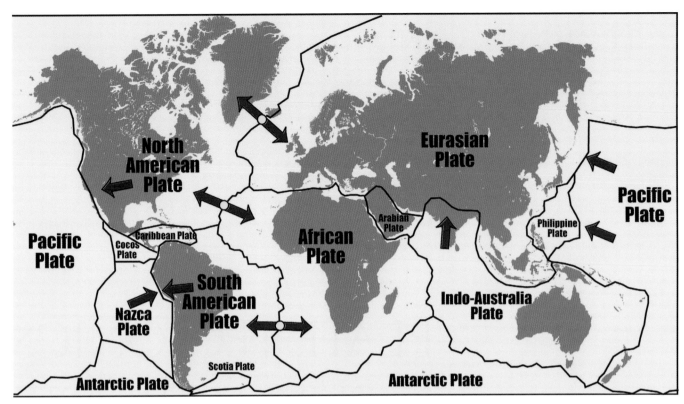

Figure 11-1. Diagram of the 12 major plates of the Earth's crust. Where arrows point away from each other, plates are splitting apart and new crust is forming. Where arrows point toward each other, plates are crashing together, such as the Nazca and South American Plates. The yellow dots sit on the Mid-Atlantic Ridge, illustrated in the cross section below.

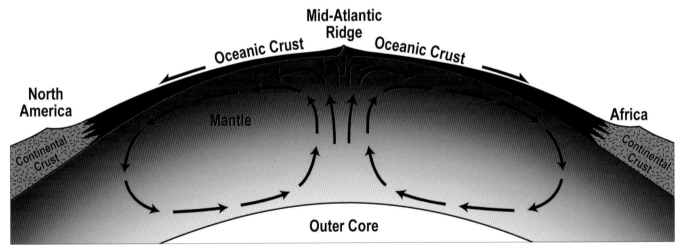

Figure 11-2. Cross section of the Earth's mantle and crustal plates, showing the convection of hot mantle rock and formation of new oceanic crust under the Mid-Atlantic Ridge.

above it. High temperatures cause several important phenomena: rock becomes more plastic at high temperatures (think warming butter), and rock expands as it heats, which lowers its density. Hotter, less-dense, semi-plastic material has a tendency to rise (slowly) toward the surface; correspondingly colder rock is denser and will sink. This behavior creates the circular movement of rock material

within the mantle that is the key mechanism driving plate tectonics.

Rates of Plate Movement

Where tectonic plates separate from each other, they move apart at a rate of a few inches per year — about the rate at which your fingernails grow. At the

Mid-Atlantic Ridge, the average rate of spreading has varied from 1 to 2 inches per year (recall example in Chapter 9). Faster rates are measured in parts of the Pacific, with rates up to about 5 inches per year. Where the plates are spreading apart at mid-ocean ridges, new oceanic crust is being formed. Vents along these ridges, called *black smokers*, pour out ultra-hot, mineral-saturated water and bear witness to the cauldron below (Fig 11-3). In places where the plates are spreading apart on land, such as Iceland, lava commonly flows to the surface (Fig 11-4).

Such cracks in the crust are where flood geologists place the "fountains of the great deep"

Figure 11-3. A "black smoker" along the Mid-Atlantic Ridge. What looks like smoke is actually minerals precipitating out when venting water hits the cold seawater. *Photo from NOAA PMEL Earth-Ocean Interactions Program.*

Figure 11-4. Basaltic lava flow in Iceland, with Northern Lights in the sky. *Photo by Sigurður Stefnisson.*

(Genesis 7:11). Supposedly, the Earth's crust was unbroken prior to the flood. Cracks in the crust released highly pressurized, mineral-rich water stored in the rocks, and thrust continental plates apart at breathtaking speeds. Current movement along plate boundaries, measured by satellites, is said to be the residual movement as plates have gradually slowed since the days of the flood. But we have already seen that this is a testable hypothesis. In Chapter 9, one of the tests for the reliability of radiometric dating also proves to be a test for the speculation that catastrophic plate motion was many times faster in the past. To recap, radiometric ages of ocean crust combined with distances between measuring points yield calculated rates for plate movement in the past that are nearly identical to the rates measured by GPS satellites today. In the north Atlantic, past rates based on radiometric dating are between 1.1 and 1.7 inches per year, which is remarkably similar to the rate of 1 inch per year measured today. This is powerful evidence against the catastrophic plate tectonics of flood geology.

DID SALTS AND LIMESTONE FORM FROM MINERALS RELEASED FROM THE "FOUNTAINS OF THE DEEP"?

Some flood geologists argue that massive limestone and salt deposits formed when hot, mineral-rich waters were released from the "fountains of the deep" as cracks in the crust opened and plates were violently thrust apart. They contend that present plate movement represents the tail-end of a process that is slowing down after a supposed catastrophic event. If so, it is reasonable to assume that the super-heated, mineral-laden water coming out of the black smokers is also left over from fluids released during the flood, and therefore the ocean floor around the vents should have at least a thin layer of salt or limestone. So what do we actually find? Lots of sulfur-rich deposits. Calcite (limestone-building mineral) is a minor constituent, and salt is entirely absent.

Earthquakes, Friction, and Plate Velocity

Earthquakes are commonplace on our Earth, and they occur primarily near plate boundaries (Fig 11-5). Because the contact between plates is not smooth, friction locks the plates together and the rock flexes where adjacent plates move in different directions. Eventually, the force exceeds the friction, and the contact ruptures (faults). Seismic waves (earthquakes) are produced as the flexed rock suddenly snaps back into position and shakes. The distribution and frequency of earthquakes provide keen insights into our dynamic planet and help provide an independent view of rates of plate motion.

The recent 2011 earthquake in Japan gives us an excellent opportunity to investigate the rate of motion. This earthquake had an overall magnitude of 9.0, the fourth strongest earthquake ever recorded. The magnitude of an earthquake is related to the distance of movement along the fault surface, how fast the movement occurred during the earthquake, and the length of the ruptured section of crust along the fault. The section of the fault surface that slipped in the Japan earthquake was 155 miles in length. Along this failed section, the greatest distance of movement

was 88 feet. The rapid movement of the crust along the ocean bottom caused tsunamis all along the eastern coast of Japan (Fig 11-6).

At first, this may seem like obvious evidence for rapid plate tectonics. But keep in mind that the stress was building at this location for hundreds of years without rupturing. When the distance of crustal slip is averaged over the time since the last rupture, the rate of plate movement is only about 3 inches per year. The slow movement of one plate under another (called *subduction*; Fig 11-7) is associated with the formation of volcanic mountains and islands. Japan and its neighboring island nations have all built up from volcanism over subducting plates, and the entire Andes mountain range in South America owes its existence to subduction of the Nazca Plate beneath the western margin of the South American plate.

If the movement of the Japan earthquake had been sustained for days or months, and combined with similar movement around the globe, as claimed by flood geologists, the heat generated by the friction would have boiled off most, if not all, of the Earth's oceans. (Recall the same expected result in Chapter 9, if radioactive decay had been much faster in the past.) To illustrate what would happen with rapidly moving plates, try a face-first,

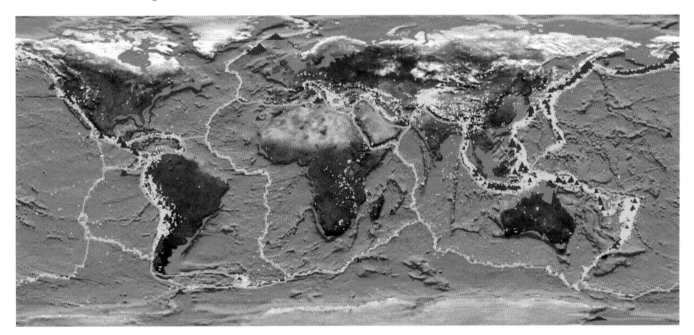

Figure 11-5. Global map with blue-green lines showing plate boundaries, red triangles showing the locations of active volcanoes, and yellow dots showing the locations of earthquakes over a 10-year period. Earthquakes are most common at plate boundaries, with the highest concentrations where plates collide. *Adapted from NASA/Goddard Space Flight Center Scientific Visualization Studio, and data from the USGS.*

Figure 11-6. Tsunami wave breaching an embankment and flowing into the city of Miyako in Iwate Prefecture, March 11, 2011. *Photo courtesy Reuters/Mainichi Shimbun.*

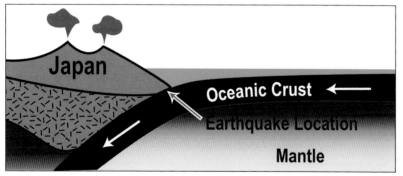

Figure 11-7. Cross section of Japan and the Pacific oceanic crustal plate. The rupture in 2011 occurred along the contact between the two plates, where the Pacific plate is being subducted beneath Japan.

home-plate slide across your carpet without a shirt on and see what happens — it's not called carpet *burn* by accident. Catastrophic plate tectonics, as described by flood geologists, is physically impossible with the laws of physics as we know them.

Rates of Mountain Uplift

Plate movement can also cause the uplift of mountain ranges. This happens where one plate is subducted under another, melting rock and generating volcanic mountains at the surface, as is the case of the Andes or Japan. It also happens where two thick continental plates collide and buckle up into mountains, such as is the case for the Himalayas (Figs 11-8, 11-9). We have already reviewed evidence proving that plate movement was never hundreds or thousands of times faster than it is now, but we will consider the possibility again in light of mountain building. What would happen if mountains were thrust up in a matter of days? Just as friction generates heat, so does flexing. Take a copper tube and bend it back and forth repeatedly, and then touch where it is bending. It gets hot! Apply the same principle to mountains at points of rapid bending, and we have the same problem – with

rapid flexing of that much rock, everything would crush and melt.

On the other hand, if we observe current rates of mountain building, such as the 1/2-inch-per-year rate in the Himalayas, and use those rates to estimate the time it should have taken to create these mountains, we arrive at time spans that are quite consistent with the time estimated for past mountain-building episodes based on radiometric dating and other geologic evidence for the timing of ancient uplift events.

CIRCUMVENTING THE HEAT PROBLEM?

A few flood geologists have suggested that supersonic water, blasting heavenward through cracks at the start of Noah's flood, would produce cooling to offset the heat of friction. Expanding gases do in fact cool, but cooling at the spot where plates rift apart would do nothing to prevent melting caused by frictional heating at subduction sites thousands of miles away. This is like saying I don't have to worry about burning my fingers on the kitchen stove because I have the air conditioner running in the bedroom.

Plate Tectonics and the Grand Canyon

Collision or separation of tectonic plates can cause portions of their associated continents to be lifted up (mountain building) or to subside (stretched or sinking downward). We'll see in the next chapter what we look for to help us determine what kind of tectonic forces were at work in the past at any given location. But for now, we'll

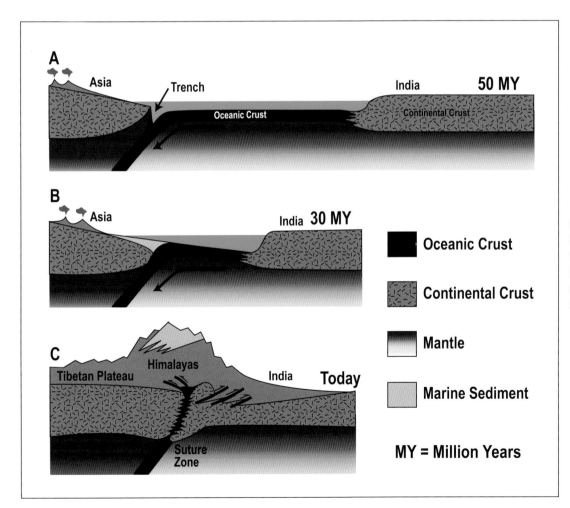

Figure 11-8. Diagram showing the collision between India and Asia and uplift of the Himalaya Mountains.

Figure 11-9. Mt. Everest, 29,035 feet – and rising! Sedimentary rock on top of Mt. Everest contains marine fossils, so before the Himalayas rose, the rock was under the ocean (as illustrated in the three-panel figure on facing page). *Photo by Hans Stieglitz.*

give a brief overview of the tectonic history of the Grand Canyon.

At least four different tectonic episodes that would have impacted the Grand Canyon region have been identified in what is now North America. Two occurred when the oldest rocks below the Great Unconformity were formed, and two happened after all the horizontal layers had been deposited and hardened. The latter two include the "Laramide episode," between about 80 and 40 million years ago, and the "Basin and Range episode," starting about 20 million years ago. The Laramide episode was caused primarily by a collision of plates, producing compressive forces that faulted and folded the Grand Canyon's layers. Much later, during the Basin and Range episode, changes in the direction of plate movement resulted in a phase of pulling apart, creating distinctly different types of faults. To figure out this history, geologists have relied in part on the study of fractures, faults, and folds – which brings us to Chapter 12.

Upturned Tapeats Sandstone in Carbon Canyon, River Mile 65. *Photo by Tim Helble*

CHAPTER 12

BROKEN AND BENT ROCK
FRACTURES, FAULTS, AND FOLDS

by **Bryan Tapp** and **Ken Wolgemuth**

History Told Through Deformed Rocks

IN THE PREVIOUS CHAPTER, WE described how the Earth's plates move as a result of the overall tectonic forces that cause earthquakes and form mountains. Now we focus on the rocks of the Grand Canyon to see what evidence of its history can be deduced from the fractures, faults, and folds in those rocks. Much can be learned from simply looking at fractures (cracks), noting the direction of movement along a fault, or observing whether a layer has been deformed by faulting or by folding. Whether a rock layer will crack, fault, or fold is determined by a combination of temperature, rock and mineral properties, rate of movement, and confining pressure.

Some types of rock are naturally more or less brittle, making some more prone to breaking and others more prone to bending. The rate of deformation is important because slower rates of motion allow rock to adjust by bending rather than break-

ing. Confining pressure refers to pressure from all sides. Think of an ice cube placed on the floor and then stepped on. We expect this ice cube to shatter. However, if the same ice cube is put at the bottom of a glacier with lots of pressure holding

WHY DOES IT MATTER WHETHER THE LAYERS WERE SOFT SEDIMENT OR HARD ROCK?

Flood geologists depend on catastrophic plate tectonics unleashed during the flood to account for all the major deformation of rock layers found around the world. The short time frame of the flood would be insufficient for the flood deposits to have hardened into rock, so flood geology presupposes that deformation had to take place while the sediments were still soft. Thus, flood geologists study the rocks to find evidence to support their presuppositions, not to see what history the rocks actually reveal.

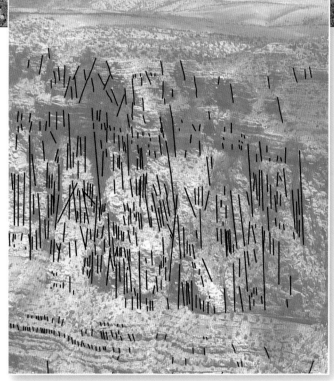

Figure 12-1. Views of the eastern wall of the Palisades, taken from the Desert View observation deck. Black lines on the inset *(right)* identify the locations of the fractures. *Photos by (left) Mike Koopsen and (right) Bryan Tapp.*

the sides in place, the ice will slowly change shape without shattering as the glacier moves. Rocks do the same thing, but the process requires very high confining pressures found only at great depths.

In this chapter, we will explore two fundamental differences between flood geology and conventional geology. Flood geologists insist that (1) all deformation of rocks above the tilted Supergroup layers occurred in soft flood sediments, and (2) most of the deformation was caused by catastrophic plate tectonics during a one-year flood.

The conventional view argues that (1) the layers were solid rock when most of the cracking, faulting, and folding took place, and (2) different types of forces were at work on the rocks (pulling them apart or pushing them together) at different times during their history.

Fortunately, this is not a matter of speculation. There are clear indicators we can look for to determine whether an ancient deposit was soft or hard when it was deformed, or whether all deformation can be explained by a single event. As an example, variable tectonic activity, such as collision or pulling apart of plates, gives rise to distinctly different types of faults. Recognizing the types of faults can thus speak to us about the

type of tectonic forces at work at the time. We can also tell much about whether layers were soft sediment or hardened rock at the time of deformation, based on the presence or absence of fractures, the size of faults, and the internal features in folds. We'll start with a look at fractures.

Fractures

Fractures are nothing more than cracks. Cracks can form in loose sediments, such as in dried clay, though they tend to be limited in size and often reseal, or "heal," when they become

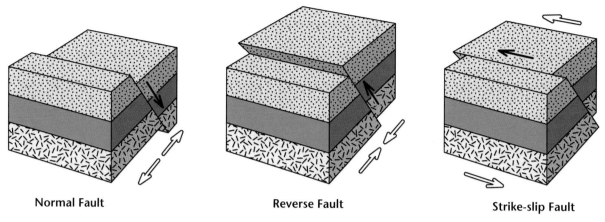

Normal Fault   **Reverse Fault**   **Strike-slip Fault**

Figure 12-2. Fault types and direction of movement.

wet. Fractures are much more common in rocks. Once they begin to form, they can grow larger with time or with increasing stress without resealing during the next rain, like mud cracks would. Long fractures that extend across multiple layers are a clear indication that all the layers were already rock before the fracture formed. When we look at the Grand Canyon, its layers are riddled with fractures that continue through formations above and below.

In the Grand Canyon, the eastern wall of the Palisades (Fig 12-1) provides a dramatic example of heavily fractured rock – rock that had become hard and brittle before tectonic stresses caused it to fracture (Fig 12-1 inset). The nearly vertical lines covering the cliff face are innumerable fractures,

spaced only 1 to 3 feet apart, with many extending through the entire cliff.

Faults

Faults are fractures in the rock where one side has moved relative to the other side. There are three main types of faults that represent three different types of stress (Fig 12-2). *Normal faults* result from extension – from rocks being pulled apart – and are recognized by a separation between layers that were once continuous. *Reverse faults* result from rocks being pushed together and are recognized by an overlap of layers that were once continuous. *Strike-slip faults* represent lateral shifts, where the two plates slide by each other sideways. Each fault type is associated

Figure 12-3. Photo of a fence that crossed the San Andreas Fault (dashed line) and was offset during the 1906 San Francisco earthquake. *Photo from the U.S. Geological Survey.*

Figure 12-4. Aerial view of the San Andreas Fault, California, looking south (strike-slip fault). The right side of the fault has moved north (laterally) relative to the left side. *Photo by Tom Bean.*

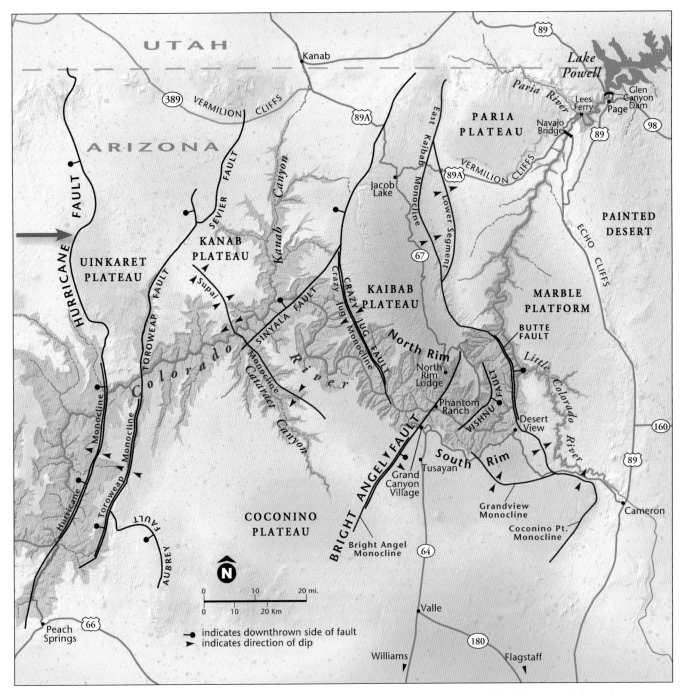

Figure 12-5. Map of the Grand Canyon showing the major faults and monoclines. Where black bar-and-ball symbols are seen along faults, the ball denotes the side that has moved down along a fault. The red arrowhead symbols along the monoclines denote the direction of downward bending. The blue arrow in the upper left shows where the photo of the Hurricane Fault was taken (discussed on page 122). *Illustration by Bronze Black.*

with very different tectonic forces and movement. Photos in this chapter illustrate some well-known faults found in the Grand Canyon and elsewhere, such as the spectacular San Andreas strike-slip fault in California (Figs 12-3, 12-4).

In the Grand Canyon, if the flood geology model were true, we should expect to see just one type of faulting. Recall that plates are said to have been violently thrust apart at the onset of the flood, with residual plate motion continuing right up to the present. With North America suddenly thrust to the west, one would reasonably expect that the American Southwest would experience compressive forces as the continent plowed into

Figure 12-6. Butte Fault at River Mile 69 with offset that reflects reverse faulting (rock on the left side of the fault sits higher than rock on the right). *Photo by Wayne Ranney.*

Figure 12-7. Outcrop showing a normal fault near the entrance to Arches National Park. The inset shows two closely spaced faults (solid lines). Layers on the right have slipped downward several feet. *Photo by Bennie W. Troxel.*

the Pacific Ocean crust, such that reverse faults should be the norm.

What we actually find in the Grand Canyon is quite different (Fig 12-5). Reverse *and* normal faults are both present, suggesting there were different periods of compressive (pushing together) and tensional (pulling apart) forces. The reverse

faults are consistent with tectonic collisions and mountain building during the Laramide episode that was mentioned in the previous chapter. The Butte Fault is a spectacular example in the Grand Canyon (Fig 12-6).

Normal faults form when rock systems experience extension – they are pulled apart. Normal

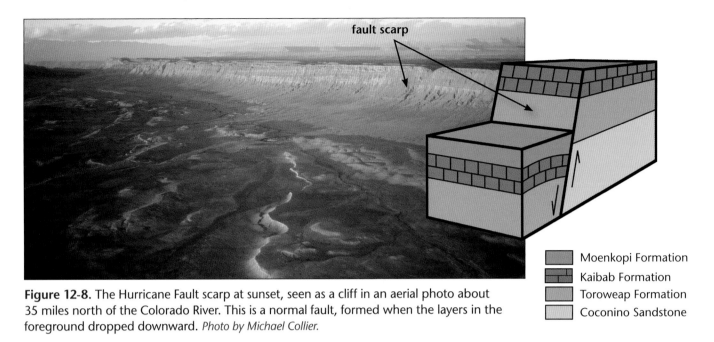

fault scarp

Moenkopi Formation
Kaibab Formation
Toroweap Formation
Coconino Sandstone

Figure 12-8. The Hurricane Fault scarp at sunset, seen as a cliff in an aerial photo about 35 miles north of the Colorado River. This is a normal fault, formed when the layers in the foreground dropped downward. *Photo by Michael Collier.*

Figure 12-9. View down the Bright Angel Fault from Yavapai Point (see map on page 120). *Photo by Mike Koopsen.*

faults are best observed on a small scale, such as found in a road cut. The lower photo on page 121 (Fig 12-7) shows a portion of the Moab fault system near the entrance to Arches National Park, north of Moab, Utah. The offset of this normal fault is clearly visible in the photo.

In the Grand Canyon, the scale of normal faults is much larger. The Hurricane Fault is a spectacular example of a large-scale normal fault, shown in the photo above (Fig 12-8). The cliff is actually the exposed surface of the fault, referred to as a *fault scarp*. The trace of the Hurricane Fault is shown on the map on page 120 (Fig 12-5), along with the location where the photo was taken.

An example of a strike-slip fault in the Grand Canyon is the Bright Angel Fault. This fault, visible

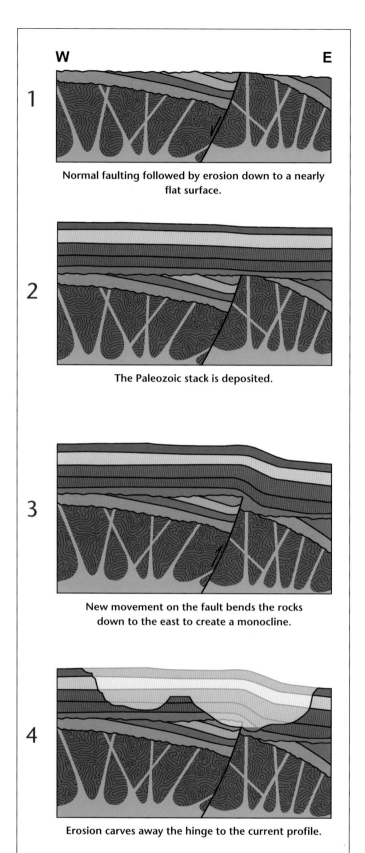

1 Normal faulting followed by erosion down to a nearly flat surface.

2 The Paleozoic stack is deposited.

3 New movement on the fault bends the rocks down to the east to create a monocline.

4 Erosion carves away the hinge to the current profile.

Figure 12-10. Sequence of a fault in the the vicinity of the Kaibab Monocline (see text for description).

from the Lookout Studio on the South Rim, forms a dramatic linear side canyon where erosion along the fault has cut deeply into the Northern Rim. Notice in Figure 12-9 that the layers on either side of the canyon stayed at the same elevation. The Bright Angel Fault has had considerable strike-slip fault movement that occurred during Precambrian time, but additionally it has also had reverse and normal fault movement during its history.

Even more significant are some faults that can only be explained if there was a time of faulting in one direction, followed by deposition of layers on top, and then later movement of the fault in the *opposite direction*. Long before the Laramide or Basin and Range episodes, during the Precambrian, the Vishnu Schist and overlying Supergroup layers were extensively faulted (pulled apart). Their sequential history is illustrated in Figure 12-10 to the left. In Frame 1, a normal fault was formed through the schist and overlying Supergroup (notice the separation of once-continuous layers), followed by erosion to a nearly flat surface to create the Great Unconformity (the wavy line across the top). In Frame 2, horizontal layers of sediments were deposited (Tapeats Sandstone and overlying layers). Frame 3 shows those horizontal layers bending to form the Kaibab Monocline. This would only have been possible if the fault changed its original downward movement on the western side (normal fault) to an upward movement as a reverse fault in order to lift the western side of the canyon. Frame 4 shows the current view that you see from the Desert View area, with portions of the overlying layers eroded away by the Colorado River.

Changes in the direction of plate movement make sense if convection cells in the mantle gradually shift with time, resulting in compressive forces during one period of time, and tensional forces during another period of time. The flood model provides no plausible mechanism for changing plate directions, since in that model all plates were set into motion by the rupturing of the crust.

Folds

Plastic deformation in rock requires a combination of high temperatures, slow rates of movement, and/or high confining pressures, which result in stretching when pulled apart, wavy folds when pushed together, and broad bending if uplift or subsidence occurs unevenly. Tight, wavy folds can be seen in some of the granite dikes in the Grand Canyon's basement rocks, but most of the folding above the Great Unconformity occurs in the form of immense bends, called *monoclines*, that can be found in the mostly horizontal rocks comprising the bulk of the layers in the canyon (Figs 12-11, 12-12).

Monoclines (literally meaning "one incline") are large-scale folds with only one upward or downward bend. A playground slide offers a visual analogy, with a single incline between horizontal ends. Monoclines occur due to regional compression of rocks and uneven uplift. There are several monoclines in the Grand Canyon region, including the Hurricane, Toroweap, Kaibab, and Supai monoclines (see Fig 12-5 map, page 120). The most spectacular of all is the Kaibab Monocline (Fig 12-12).

Figure 12-11. Folds in the Vishnu Schist. The folding happened in the metamorphic rock (schist) while it was still hot and easily deformed. *Photo by Wayne Ranney.*

Figure 12-12. The Kaibab Monocline, looking north-northwest. See panorama, pages 238-239. *Photo by Tom Bean.*

Figure 12-13. Compressional folding in the Tapeats Sandstone, Carbon Creek area. The faded photo to the right shows traces of some layers, some of the fractures, and the changing direction of the folds resulting from flexural slippage. Note the similarities to the lower diagram of Figure 12-14 below. *Photo by U.S. Geological Survey.*

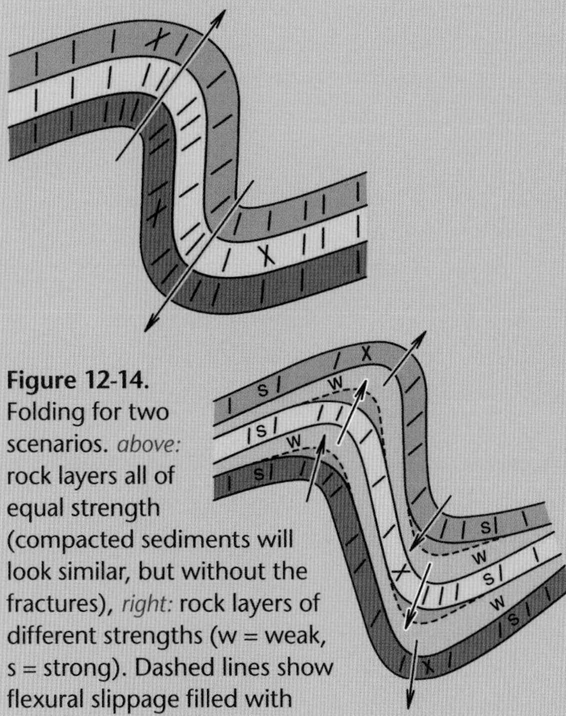

Figure 12-14. Folding for two scenarios. *above:* rock layers all of equal strength (compacted sediments will look similar, but without the fractures), *right:* rock layers of different strengths (w = weak, s = strong). Dashed lines show flexural slippage filled with weaker rock.

The origin of this giant fold is illustrated in Frame 3 of the sequential diagram on page 123, where uplift on just the western end of the canyon created the Kaibab uplift, and once-horizontal beds now bend downward to the east.

A number of smaller folds are associated with regional faults, where rock has been twisted and bent along a fault. The most spectacular of these are located along the Butte Fault — with select folds often cited by flood geologists to argue for deformation while the sediment was still soft (Fig 12-13, also see photo on page 126). To identify folding in sediment versus rock, we need some background on what to look for.

In rock layers, bending results in numerous fractures in each layer that do not heal (reseal). A greater number of fractures are expected in the stronger, more brittle layers. Folding also creates stresses between layers that result in one layer slipping forward or backward relative to the underlying layer (Fig 12-14, lower diagram). This phenomenon, referred to as *flexural slippage*, creates a gap that may be filled in later with weathered material, or weaker rock may deform into the space. Such slippage can be identified by looking at the direction the fold "points" in each layer. The sharpest part of the fold (the "hinge," marked with arrows) points in a different direction in each layer, due to slippage between layers.

CAN FAULTS FORM IN WET, SOFT, AND PLIABLE SEDIMENT?

Faults in rock look very different from those in soft sediment. In rock, a relatively clean break occurs that is often filled with angular fragments of broken rock (called *breccia*), or with pulverized rock as one side of the fault grinds past the other. Where bending of nearby rock occurs, cracks are readily visible in the deformed rock. In soft sediments, however, there is no clean break and sediments are spread out along the blurred rupture zone. Because the material is soft, there is little or no breccia, nor do we find pulverized material lining the sides. Faults in the Grand Canyon are characterized by sharp breaks filled with rock fragments, and bent layers adjacent to faults are fractured. Faults in soft sediment don't look like this.

In contrast, when sediment is folded, the soft material readily crumbles to fill in cracks, resulting in few, if any, unhealed fractures. Grains within each layer shift as the sediments are compressed, greatly reducing stress between layers. This means we don't expect to see flexural slippage between layers of uncemented sediment. With no flexural slippage, the fold hinges in consecutive layers all point in the same direction (Fig 12-14, upper diagram).

Revisiting the folding in the Tapeats Sandstone next to the Butte Fault, we find both flexural slip *and* abundant fractures (Fig 12-15). Flood geology descriptions claim that folds at this location are not fractured, and accompanying photos often do look un-fractured. With sufficient photo resolution, however, or with a trip to visit the fold personally, the fractures can be easily identified.

below: **Figure 12-15.** Close-up of part of the Tapeats fold in Carbon Canyon, in the area of the sharpest bend near the middle of the lower red arrow shown in Figure 12-13. A slip between layers is highlighted in the inset.

WHAT WOULD FOLDS IN FRESHLY DEPOSITED FLOOD SEDIMENTS LOOK LIKE?

Flood geologists claim that sediment layers can form folds that look like those found in the Grand Canyon, but there is more to the story than just *soft* versus *hard*. It is also critical to consider how recently the sediment layers formed. Folding that retains the general layering in un-cemented sediments is possible only if those sediments have had time to settle and compact. Sediments freshly deposited in lakes or in the ocean have a high water content that make them comparable to a cold milkshake. Imagine a milkshake with alternating layers of vanilla and chocolate. As long as the milkshake is kept cold and undisturbed, the layers stay intact. But if it is shaken or tipped over, the layers lose their consistency and flow into mixed blobs of white and brown. Freshly deposited sediments from a global flood would be very similar. As soon as folding began, layers would have quickly lost their consistency and flowed into a jumbled mess. This isn't just speculation — we occasionally find isolated layers preserving just such a history, such as in the example below (Fig 12-16).

Figure 12-16. Dewatering and soft sediment deformation in the Nellie Bly Formation, west of Tulsa, Oklahoma. *Photo by Stephen Moshier.*

This wraps up our general discussion of Part 2.

This section — How Geology Works — was intended to give you a better understanding of what geologists look for in determining what happened at any particular location in the past. It also explains how they assess the competing claims for Earth's physical history. In the next section, we will shift our focus to past life on Earth — to fossils. Do the fossils in the Grand Canyon and Grand Staircase reflect the Young Earth claim that a global flood is the most logical explanation for the trillions of deceased organisms entombed in these rocks? Or, do they reflect normal Earth processes over geologic time?

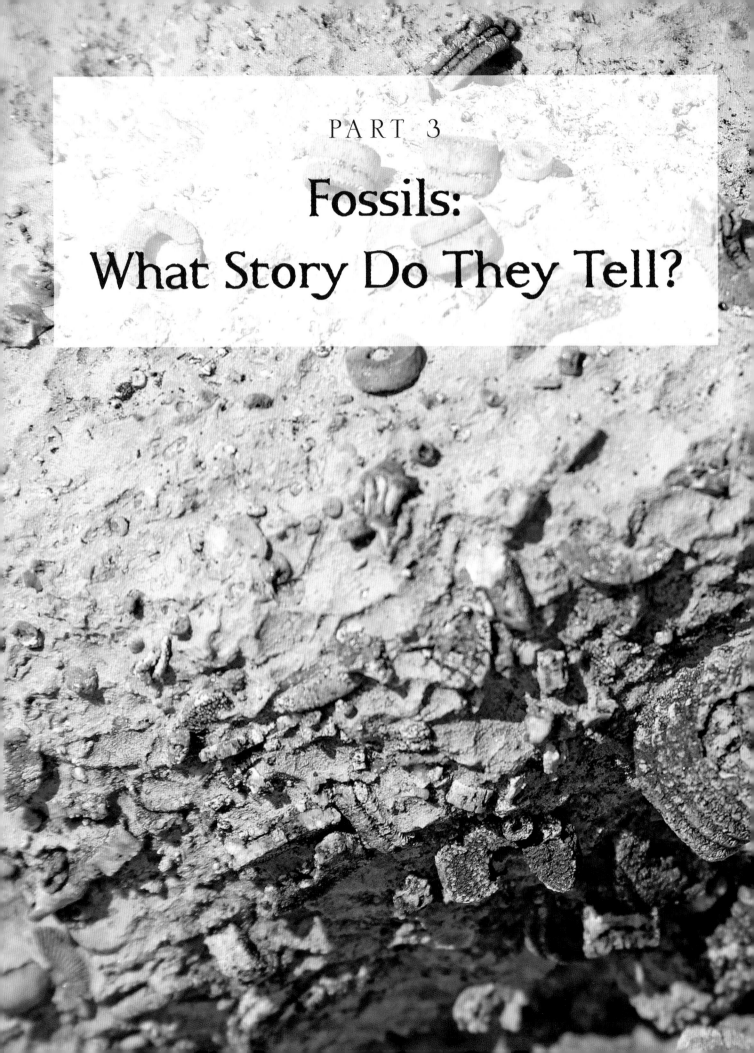

PART 3

Fossils:
What Story Do They Tell?

In Part 2, we addressed fossils in general terms as a dating tool. Once it was recognized that rock layers around the world contain a consistent ordering of fossil organisms, it became possible to use the presence of specific fossils to identify the particular interval of time represented by a given rock layer. Recall that this was not based on any assumption of organisms becoming increasingly complex over time, but instead on the simple observation that particular suites of organisms abundant in one layer are replaced by different suites of organisms in overlying layers.

In Part 3, fossils get center stage, with attention paid to the particular story revealed by ancient organisms in the Grand Canyon. Before diving into this subject, we'll briefly revisit Chapter 2 in order to remind ourselves why we are doing this. Recall that the Young Earth view claims that Noah's flood covered the entire planet and that no organisms died before the fall of Adam (prior to the first sin). From this theological position, it is inferred that all of the death represented by the fossils in the Earth's layers must be predominantly from Noah's flood, which forces the conclusion that the events surrounding the flood must have been monumentally violent. To flood geologists, fossils represent the "all flesh" of Genesis 7:19-20 that perished in a year-long global flood. Animals and plants were carried to the Grand Canyon area from distant locations by powerful tsunami-like currents and then were quickly buried by sediment. And, because the stack of Grand Canyon and Grand Staircase sedimentary rocks is collectively no less than two miles thick (and as much as three miles thick), the proposed deluge must have been correspondingly deep and violent.

But do fossils support this view? The modern geology view is that fossils represent the remains of animals and plants — some from hundreds of millions of years ago — that lived, died, and were buried, most in close proximity to where they lived.

Drawing on the particular expertise of three different paleontologists (scientists who study fossils), the next three chapters focus on the fossil record found in the Grand Canyon and Grand Staircase rocks, and how it compares to the global fossil record. The first chapter (Chapter 13) addresses animal fossils. The second chapter (Chapter 14) considers plant fossils, with special attention to what can be learned by studying microscopic spores and pollen. The third chapter (Chapter 15) focuses on trace fossils (things like fossilized tracks and burrows), particularly as they relate to the deposition of the Coconino Sandstone.

Crinoid fossils in the Redwall Limestone. *Photo by Bronze Black.*

Fossil corals and bryozoans in the Redwall Limestone. *Photo by Bronze Black.*

CHAPTER 13

FOSSILS OF THE GRAND CANYON AND GRAND STAIRCASE

by **Ralph Stearley**

Faunal Succession: A Global Perspective

W E'LL START THIS CHAPTER WITH a recap of the Principle of Faunal Succession, which was described in Chapter 8. Those who first proposed this principle did so primarily on the basis of observations of different marine shelled organisms in successive layers, with little attention or notice of changes in organism complexity from one layer to the next. As more fossil studies were conducted, however, it also became apparent that younger layers contain fossils of increasingly more complex or more diversified organisms.

The oldest fossil-bearing rocks on Earth contain the remains of the simplest organisms – bacteria and algae – whereas overlying layers contain organisms with more complex body plans. Still younger layers contain fossils of fish and other marine creatures, but no evidence of any land-dwelling organisms. Land dwellers make their first appearance in even younger layers. Layers above the first land animals and plants eventually contain reptiles, dinosaurs,

and, ultimately, birds, mammals, and flowering plants. Some types of organisms that first appear in ancient rock layers of a particular age continue to be present in younger layers up to the present, while others vanish from the rock record in younger layers. (This vanishing act, when species disappear from the fossil record, is called *extinction*).

As an example, fossil sharks of various types can be found in all layers after their first appearance in the middle Paleozoic Era, but large marine reptiles that once ruled the seas first appear in Mesozoic layers and then disappear from younger layers. Many categories of organisms are *never* found mixed together; for example, marine reptiles like ichthyosaurs are not found with whales or dolphins, and trilobites are never found with sand dollars or penguins.

The fact that some types of fossils are never found together is our first clue, on a global scale, that fossils do not represent animals that were buried in Noah's flood. There is an order to the fossil record, both in complexity and in the kinds of fossils found. Young Earth advocates generally acknowledge this order but disagree on how it came to be. Some argue

Era	Geologic Period	Examples from the Faunal Fossil Record in Grand Canyon and Grand Staircase Rocks
Ceno-zoic	Paleogene and Neogene	**Terrestrial:** freshwater snails (Claron Fm); freshwater turtles, fragmentary mammal remains (Brian Head Fm).
Mesozoic	Cretaceous	**Marine:** ammonites, sharks, plesiosaurs (Tropic Shale); oyster reefs, snails, clams. **Terrestrial:** frogs, lizards, turtles, crocodiles, duckbill dinosaurs, horned dinosaurs, theropod dinosaurs, marsupial mammals (Kaiparowitz and Straight Cliffs Fms).
Mesozoic	Jurassic	**Marine:** clams, snails, crinoids, ammonites (Carmel Fm). **Terrestrial:** dinosaur bones and tracks (Entrada, Navajo and Kayenta Fms).
Mesozoic	Triassic	**Terrestrial** (mostly): freshwater clams, crayfish, beetles, lungfish, phytosaurs, aetosaurs, early bipedal dinosaur, *Coelophysis* (Chinle Fm); tracks, large amphibians, coelacanth fish (Moenkopi Fm).
Paleozoic	Permian	**Marine:** sponges, corals, brachiopods, bryozoans, gastropods, pelecypods, trilobites (rare), nautiloids, sharks, bony fish (Kaibab and Toroweap Fms). **Terrestrial:** tracks of reptiles, spiders, and scorpions (Coconino Ss).
Paleozoic	Mississippian and Pennsylvanian	**Marine:** brachiopods, gastropods, pelecypods, corals, bryozoans, conodonts, sharks (Supai Gp, Redwall Ls, Surprise Canyon Fms).
Paleozoic	Devonian	**Marine:** massive stromatoporoid sponges, rugose corals, tabulate corals, brachiopods, trilobites, placoderm fish, conodonts (Temple Butte Fm).
Paleozoic	Cambrian	**Marine:** early trilobites, brachiopods, echinoderms, sponges (Muav and Bright Angel Fms).
Proterozoic	Late Proterozoic	No animal or multicellular fossils in Grand Canyon Supergroup.
Proterozoic	Early Proterozoic	No fossils in crystalline basement rock.

Figure 13-1. Summary of faunal fossil record in Grand Canyon/Grand Staircase and global layers. Blue text indicates examples specifically from the Grand Canyon/Grand Staircase.

Taxonomic groups appearing:
Marine: whales, dolphins, sand dollars.
Terrestrial: almost all current mammals and birds.

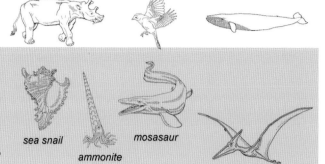

End-Cretaceous extinction: taxonomic groups
completely disappearing: ammonites, *all* dinosaurs
(except birds), pterosaurs, mosasaurs, ichthyosaurs,
plesiosaurs.

Taxonomic groups appearing: neogastropods, snakes,
mosasaurs.

sea snail *mosasaur*
ammonite
pterosaur

Taxonomic groups appearing:
Marine: entirely new plankton groups, including planktonic
foraminifera and diatoms, plesiosaurs, rays (including
stingrays), most familiar shark groups.
Terrestrial: butterflies, moths, first birds, lizards, giant
sauropod dinosaurs.

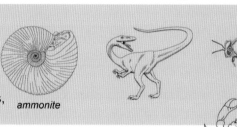

oyster

End-Triassic extinction: many marine groups including
conodonts, and many terrestrial tetrapod groups
disappear.

Taxonomic groups appearing: **Marine:** modern
scleractinian corals, oysters, decapod crustaceans like
crabs and lobsters, ichthyosaurs. **Terrestrial:** ants, bees,
wasps, earliest dinosaurs.

ammonite

End-Permian extinction: 95% of marine invertebrates go
extinct. Taxonomic groups **completely disappearing:**
Marine: *all* trilobites, rugose corals, cryptostome and
fenestrate bryozoans, blastoid echinoderms, fusulinid
foraminifers, orthid, spiriferid, and productid
brachiopods.

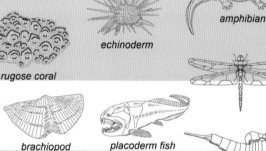

bony fish
mammal-like reptile
amphibian

Taxonomic groups appearing:
Terrestrial: centipedes, spiders, and many early
amphibians.

echinoderm
rugose coral

Late-Devonian extinction: Groups that **disappear:**
Marine: brachiopods (orders Atrypida and Pentamerida);
placoderm fish. (Note: Taxonomic groups appearing in
Ordovician, but not found in Grand Canyon: Phylum
bryozoan, nautiloid cephalopods, and rugose corals.)

brachiopod *placoderm fish*

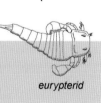

Taxonomic groups appearing: **Marine only:** most major
invertebrate phyla appear; no known terrestrial biotas.

eurypterid

sponge *trilobite*

Taxonomic groups appearing in the latest Proterozoic:
Phylum Porifera (sponges) *(Paleophragmodictya),*
Phylum Cnidaria (including conulariids), flatworms,
Phylum Echinodermata, monoplacophoran mollusks
(Kimberella), the earliest arthropods *(Parvancorina).*

kimberella *Paleophragmodictya Parvancorina*

that more-complex organisms could flee to higher ground, that earthquake activity caused vibrations that sorted organisms by size, or that different biological communities were swept from different locations into stacks that produce the observed order. Yet none of these mechanisms can explain why fossils of a marine mammal and a marine reptile (both air breathers) that lived in the same kind of environment are never mixed, or why the marine reptiles are found only in layers older than those containing mammals. If colossal walls of water surged over entire continents, why do we not see mixtures of marine and land organisms *as the norm?*

Faunal Succession in the Grand Canyon and Grand Staircase

So what do we find on a more localized scale in the Grand Canyon and surrounding region? The table on the preceding two pages summarizes the fossils contained in Grand Canyon and Grand Staircase rocks – from the lowest sedimentary layers in the Grand Canyon up to the highest layers in Bryce Canyon in Utah (Fig 13-1). It also includes some global examples of fossils found in other rocks of the same age from around the world. The purpose of this table is not to overwhelm with facts or to impress with big words. It is included because it illustrates what paleontologists have actually found in the Earth's layers. There is an amazing order in the fossil record!

The most important aspects of this table:

1. This table of fossils is not hypothetical; it's what is simply there. Anyone can go and observe the fossils in these rock formations. Thousands of non-professional rock hounds do so every year. Proponents of either viewpoint (flood geology or modern geology) must explain this sequence of fossils in order for their claims to be considered valid.

2. The increase in complexity and diversity from the lowest (oldest) to the highest (youngest) rocks is readily apparent. In Precambrian rocks only simple algae are found, but as one goes up the entire sequence, the age and character of the fossils change from dominance by invertebrates (Paleozo-

WHAT IS THE CAMBRIAN EXPLOSION?

The presence of fossils with complex body plans and hard shells in Cambrian rock is often referred to as the "Cambrian Explosion" because of the apparently sudden appearance of these life forms with few obvious predecessors. Young Earth advocates proclaim this as clear evidence that complex life started all at once at this juncture. But two pieces of information are left out. First, many fossil life forms of simpler design are known from rock layers older than the Cambrian, and older layers also contain many preserved tracks and burrows. Second, the presence of hard shells hugely increases the likelihood of preservation as a fossil. In other words, during Earth's early history when organisms lacked shells, few deceased bodies were preserved. Once shells appeared, fossilization became more common – resulting in exactly what we see, a seemingly abrupt proliferation of these fossil types.

ic), to the appearance and proliferation of dinosaurs (Mesozoic), to the appearance and proliferation of mammals and birds (Cenozoic).

3. This table also denotes the major extinctions that have occurred over the history of life on planet Earth. After each extinction, fossils from major taxonomic groups appear or disappear in the Grand Canyon-Grand Staircase sequence with the same order and timing as observed globally. This is a conspicuous trend in the history of life, and, in fact, boundaries between several geologic eras or periods are defined by the disappearance of one or more types of life forms.

Grand Canyon Fossils Compared with Modern-Day Planet-Wide Biotas

If we follow the Colorado River downstream and on into the sea, what kinds of living creatures will we find? The shallow waters of the Gulf of California, not far from the Grand Canyon, host many thousands of species visible to the naked eye. Inhabitants of the shallow seafloor include an amazing variety of creatures such as crabs, lobsters, shrimp, barnacles, sand dollars, octopi, stingrays, flounder, and halibut. The gulf's open waters teem with tuna, swordfish, groupers, sea turtles, dolphins, sea lions, and whales. Seagulls and pelicans skim the surface and dive beneath waves. Moving to colder waters farther south, we can add elephant seals, penguins, and krill. The list swells immensely if we consider microscopic organisms, with myriad unique species of shell-producing creatures, including distinctive diatoms, forams, and coccolithophorid algae. If we look for fossil examples of these creatures in the Grand Canyon rock sequence, how many do we find? The answer is not just a low number — it is *zero*.

We do find corals, brachiopods, and snails both in modern seas and in Grand Canyon layers, but we discover something fascinating when we look closely at the shapes and specific characteristics of each species — the modern ones look totally different from those in the Grand Canyon. Some typical fossils from Grand Canyon strata (and equivalent strata extending into Nevada) are identified on pages 137-139 (Figs 13-5 through 13-14).

Biologists classify organisms using a naming system in which the relatedness of organisms decreases as we move from species, to genus, to family, and on through order, class, phylum, kingdom, and domain. If we look at coral or snail fossils in the Grand Canyon, for example, they are not just different from modern varieties at the species or genus level, they are so distinct that biologists place them in different orders. Placoderm fish, found in the Temple Butte Formation, are so distant in form and function from modern fish that they are not even considered to be in the same class. And the many species of trilobites found in the canyon rocks are not present in any form in today's oceans.

To illustrate how large the differences can be at the order and class level, consider the example in the box on biological classification. Dogs share the same *order* with creatures as diverse as tigers, polar bears, and weasels, and the same *class* as mice, whales, and kangaroos (see box below).

BIOLOGICAL CLASSIFICATION OF ORGANISMS

Biologists today utilize a classification system that organizes living creatures according to shared characteristics. The Linnaean system, named after the 18th century natural historian Carolus Linnaeus, groups organisms into a series of categories of increasing similarity (Fig 13-2). At the highest level, called a *domain,* the shared characteristic may be as broad as having a cell nucleus and including organisms as diverse as mushrooms and monkeys. With each subsequent category, the organisms included become more and more similar. At the bottom, organisms of the same interbreeding kind are referred to as *species*. Scientific names are based on the genus and species. An example is shown to the right for classification of the dog, *Canis familiaris.*

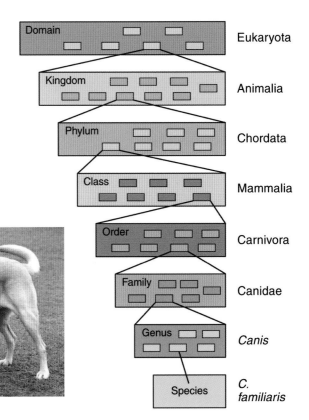

Figure 13-2. Linnaean classification system for organisms. *Photo by D.J. Mirko.*

shots of life's diversity in stacked layers of sediment and rock.

This situation – global faunal replacement, in which one group of organisms replaces another – should be extremely puzzling to someone who believes that all this is the result of a single, catastrophic flood in the recent past. Why are almost all of the groups of creatures that are preserved as fossils in the strata of the Grand Canyon now extinct? Why can't we locate any representatives of the modern marine fauna listed above in the Grand Canyon strata, if Earth is supposedly only a few thousand years old? And, as we've asked before, why didn't massive ocean waves barreling over the surface of continents thoroughly mix marine and terrestrial organisms into the same deposits?

The Bible itself speaks against Noah's flood as an explanation for the observations above. Genesis 1 lists major taxonomic groups of plants and animals in the pre-flood creation that match those alive today: seed-bearing plants and trees that bear fruit with seed in it (v. 11), birds (v. 20), and livestock such as cattle and wild animals (v. 24). Why aren't any of these modern creatures present in the layers of the Grand Canyon, which should be a record of life that existed just before the flood?

But also recall that colossal waves (tsunamis) are said by flood geologists to have hurtled across entire continents, so all manner of terrestrial (land) life forms should be mixed in with marine deposits. Our continents are inhabited by great numbers of snakes, turtles, moles, rats, gazelles, wolves, kangaroos, sparrows, parrots, hawks, monkeys, and apes, as well as freshwater trout, bass, piranha, crayfish, and alligators. To this list we can add the incredible assortment of modern plants, dominated today by flowering varieties (angiosperms), including oak and maple trees, a plethora of wildflowers, and all our beans, fruits, nuts, and vegetables. How many of these are found in the Grand Canyon layers? Again, it is not just a low number – it is zero.

The simplest explanation is that these modern types of plants, invertebrates, fish, reptiles, amphibians, mammals, and birds simply did not exist during the time when the Grand Canyon sediments were being deposited. Different life forms came and went over the Earth's history, leaving behind snap-

Life through Time in the Grand Canyon Region

The fossils actually found in the Grand Canyon and Grand Staircase tell a story of changing life forms over time that matches similar changes observed all over the Earth. The descriptions of fossil life in the pages that follow are summarized in the table on pages 132-133 (Fig 13-1).

Precambrian Life

Precambrian rocks are represented by the huge stack of tilted sedimentary rocks under the Great Unconformity (the Supergroup). Some of these contain fossils. The flood geology authors of *Grand Canyon: A Different View* suspect that these

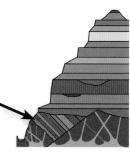

fossils were tilted along with their strata during the early phase of the great flood catastrophe — and thus, they represent a "pre-flood" biological community.

Stromatolite mound

From the perspective of a flood geologist, all the major categories of organisms living today were present prior to the flood, so the pre-flood layers should contain an assortment of all these life forms. Yet the only evidence of fossil life — in *thousands* of feet of accumulated Supergroup layers — are structures produced by colonial algae (stromatolites, oncolites), found in the Bass Formation and elsewhere in the Supergroup (Figs 13-3, 13-4). Stromatolites (and the smaller, free-rolling oncolites) are structures produced as sticky films of single-celled algae that accumulate in muds. No multicellular organisms are found anywhere in the Supergroup, even though the thickness equals that of the layers above that are said to have been laid down by the flood.

Interestingly, the Supergroup layers contain fossils of several additional single-celled organisms paleontologists recognize as planktonic algae (*plank-*

tonic refers to floating or drifting). How is it that the ancient seafloor-dwelling stromatolites were buried along with floating planktonic algae — but with no other floating organisms or any multicellular organisms at all — when Genesis 1 states that there were multicellular plants and animals (including fish) that existed in pre-flood days? This not only violates a literal reading of the Bible, it also violates common sense.

Paleozoic Life

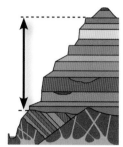

The Paleozoic Era is represented in the Grand Canyon by layers sitting on top of the Great Unconformity: from the Tapeats Sandstone up to the Kaibab Formation. Globally, trilobites are located only in Paleozoic strata (Fig 13-5). Within this window of time, we can clearly differentiate between earlier and later trilobite faunas. For example, the small, eyeless trilobites of the order Agnostida are easily recognized and common in Cambrian and Ordovician layers worldwide, but are completely absent from layers after the Ordovician.

Rolled phacopid trilobite

On the other hand, trilobites belonging to the family Phacopida, which could roll themselves up into protective balls, are common in Ordovician through Mississippian rocks, but are completely absent from the older Cambrian rocks.

The Cambrian rocks of the Grand Canyon (Tapeats Sandstone, Bright Angel Shale, and Muav Limestone), like all Cambrian rocks around the

Figure 13-3. Stromatolites, Chuar Group. *Photo by Wayne Ranney.*

Figure 13-4. Rock face exposing a cross section of oncolites (similar to stromatolites), Chuar Group. *Photo by Stephen Threlkeld.*

Figure 13-5. Trilobite *Dolichometopus productus* from the Bright Angel Shale. *Photo courtesy Grand Canyon National Park GRCA #17187.*

Figure 13-6. A Devonian placoderm, *Dunkleosteus*, along with other marine creatures of that time. Smithsonian Institution mural, Washington, D.C. *Photo by Tim Helble.*

Figure 13-7. Head of an antiarch placoderm *Bothriolepis* as viewed from above, from a Devonian formation in Quebec, Canada. The same genus is widespread and has been found in the Devonian Temple Butte Formation in the Grand Canyon. *Photo by David Elliott.*

Figure 13-8. Productid brachiopod, Redwall Limestone. *Photo by Debbie Buecher.*

Figure 13-9. Crinoid remains with some disarticulated pieces, Redwall Limestone. *Photo by Howard Lee.*

Figure 13-10. Fossil coral, order Tabulata, *Syringopora*, Redwall Limestone. *Photo by Debbie Buecher.*

Figure 13-11. Bryozoan from the Redwall Limestone. *Photo by Michael Quinn, courtesy Grand Canyon National Park.*

Figure 13-12. Dragonfly wing, Hermit Formation. *Photo courtesy Grand Canyon National Park, GRCA #3090.*

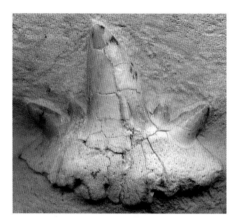

Figure 13-13. Shark tooth, Kaibab Formation. *Photo by David Elliott.*

Figure 13-14. Fossil horn coral, order Rugosa, Fossil Mountain Member, Kaibab Formation. *Photo by Bob Buecher.*

below: **Figure 13-15.** Diorama depicting life existing at the time the Mesozoic Chinle Formation was being deposited. *Courtesy Petrified Forest National Park.*

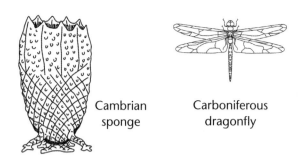

Cambrian sponge

Carboniferous dragonfly

world, do not host any corals or bryozoans. Rock units deposited above the Cambrian *do* host corals and bryozoans (photos on the facing page), all or most of which belong to totally extinct orders.

There are no bony-fish remains of any kind in the Cambrian or Precambrian rocks of the Grand Canyon. The first bony-fish fossils we find are in the Devonian Temple Butte Formation, but they are the remains of placoderms, a unique group of extinct vertebrates that are common globally in Devonian rocks, but then disappear from the rock record entirely (Figs 13-6, 13-7). Throughout the Grand Canyon strata, there are many similar patterns of upward faunal replacement of large taxonomic groups within the sponges, brachiopods, echinoderms, and molluscs, as well as other kinds of creatures (Figs 13-8 through 13-14).

Mesozoic Life

But wait, there's more! Above the rim of the Grand Canyon lies a large stack of layered rocks, which flood geologists claim were

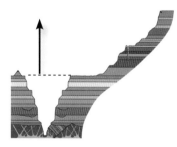

also deposited during Noah's flood. These form the "Grand Staircase" series of rocks located north of the Grand Canyon, all the way to Bryce Canyon (see Fig 3-2, pages 32-33). These sedimentary rocks,

Figure 13-16. Partial section of a diorama of Cretaceous life, recreated from fossils found in Grand Staircase rocks. Note the pine (conifer trees) in this diorama. At this time there were no flowering-type trees (see next chapter). *Photo by Wayne Ranney. Mural created by Larry Felder for Big Water Visitor Center, Grand Staircase-Escalante National Monument, Utah (used with permission).*

beginning with the Moenkopi Formation and extending up to the Claron Formation, are considered Mesozoic and early Cenozoic by modern geologists. What sorts of fossils do these rocks contain?

Many of the Mesozoic layers in the Grand Staircase contain only terrestrial fossils (land and freshwater organisms). One example of a terrestrial deposit is the Chinle Formation. The Chinle Formation consists of colorful mudstones and siltstones that erode to form the low, rounded hills of Arizona's Painted Desert and Petrified Forest National Park. Fossils preserved there include many primitive conifers, plus several groups of plants that are unknown from underlying Paleozoic rocks. In addition, there

are insects and scorpions, some freshwater fish, and a variety of tetrapods (four-legged vertebrates), including early dinosaurs (Fig 13-15). There are no marine organisms of any kind in the Chinle, which is baffling if one holds to a scenario of these rocks having been deposited in the "late flood period" of Noah's flood.

Other stratigraphic formations in the Grand Staircase, such as the Cretaceous-age Kaiparowits Formation and the Jurassic-age Navajo Sandstone, contain dinosaur footprints, eggshell fragments, and bones (Figs 13-16, 13-17, 13-18). Again, if Noah's flood was global

Figure 13-17. Excavating hadrosaur bones from the 75 million year old Kaiparowits Formation. *Photo by Alan Titus.*

Figure 13-18. Broken dinosaur eggshells in the Kaiparowits Formation, Grand Staircase-Escalante National Monument, Utah. *Photos by Billy Doran.*

Gryphaea (oyster)

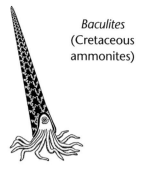

Baculites
(Cretaceous
ammonites)

as claimed, why are there no dinosaur remains incorporated into the layers of the Grand Canyon? Moreover, why are there intact dinosaur footprint trails in multiple layers of the higher Grand Staircase rocks, since footprints should not have been preserved at any stage of a raging flood? And, how could dinosaurs have laid their eggs in nests anywhere on planet Earth months into a global, catastrophic flood with tsunamis sweeping the continents from coast to coast?

Still other formations in the Grand Staircase, such as the Carmel Formation, do contain marine faunas. These rocks host many marine invertebrates, including corals belonging to the order Scleractinia, oysters, many other bivalves, and the extinct cephalopods known as ammonites.

The Mancos Shale, a Cretaceous rock unit north of the Grand Canyon, but not in the Grand Staircase series of rocks, hosts large marine reptiles known as plesiosaurs, which to date have been located only in rock layers of Mesozoic age. Again, why are there plesiosaurs but no whales, dolphins, or sea lions preserved as fossils in the Mancos Shale? Why are none of these large marine vertebrates found in the rocks of the Grand Canyon? The simplest explanation is that these animals did not yet exist when the Grand Canyon sediments were forming.

All of this specific information confirms the Principle of Faunal Succession: fossil organisms (both fauna and flora) succeed one another in a definite and recognizable order, and the overall character of life in each geologic formation differs from that in the formations above and below it. In other words, the fossil record of the Grand Canyon does not contain a hodge-podge of organisms, as would be expected if its rocks were deposited in a tumultuous global flood.

HALF A STORY TOLD

Young Earth advocates attempt to force the fossil record to agree with their claim that a year-long catastrophe deposited all the layers from the top of the Precambrian through most of the Grand Staircase. To do this, they must minimize or ignore the notion of *biotic* succession in the rock record which is explained in this chapter. For example, "corals" are mentioned as common today and in Grand Canyon fossils, without explaining that the Grand Canyon layers host *extinct* groups of corals and that modern corals are *completely absent* from Grand Canyon strata. Thus, only a part (and a small part) of the whole picture is presented.

Another tactic argues that complex organisms like trilobites are found in the lowest Grand Canyon strata – thus complex organisms appear very early in the record, while simple organisms like sponges appear higher in the Grand Canyon strata. This type of argument superficially seems to demonstrate that complicated organisms appear earlier than simple organisms; but again these observations need to be placed into the larger picture. While it is true that trilobites appear early in the record, they *completely disappear* from the rock record globally following the Permian. Meanwhile, *many* other complex organisms like bumblebees, lobsters, stingrays, dinosaurs, ducks and dolphins enter the rock record in particular layers *above those in the Grand Canyon*, as described in this chapter. Some of these, like the dinosaurs, go extinct along the way.

And while simple organisms such as sponges are found in many Grand Staircase layers – and are common today – they are very *different* from one time period to another. For example, the group of sponges termed Archeocyaths appear only in Cambrian rocks globally and then disappear from the record. Moreover, sponge fossils are documented well down into the Precambrian and globally appear (as predicted) much earlier than trilobites or other complex organisms. Thus, again, much of the biological and stratigraphical context of these fossils is ignored or unexplained in the Young Earth literature – definitely a "story half told" – only relating those parts that superficially fit with the flood geology model.

Catastrophic Transport and Hydrodynamic Sorting?

Flood geologists claim that they see evidence in the canyon of catastrophic transport and of sorting of fossil organisms by various hydrodynamic processes. Some fossils do indeed show evidence of transport, though not as the norm. For example, the fossil logs of the Petrified Forest were obviously moved – most likely rolled. Their major limbs have been broken off. However, the fossil logs are found in association with preserved remains of delicate organisms such as fern fronds, meaning that these deposits cannot have formed from a single ultra-violent event. Many of the delicate marine creatures of the Grand Canyon have been disarticulated (fallen apart), but the dissociated parts are not widely scattered from one another, nor do they appear scratched or worn down by tumbling.

Moreover, these fossils are often found in recognizable associations, meaning they are grouped with other fossil organisms typical of a particular environment, such as a shallow seafloor or lowland swampy forest. In other words, if these organisms were transported, they were not moved far from the original environment in which they lived. For example, corals thrive in shallow tropical marine waters. Grand Canyon strata with corals also contain creatures like bryozoans that lived attached to the seafloor in similar marine environments.

Other flood geologists have called upon sorting action by earthquakes to separate deceased organisms on the basis of size or density. But fossils show no preferential distribution by size or density. Abundant microscopic fossils are dispersed widely in fossil-bearing layers, and small, shelled organisms routinely appear in layers above and below layers with larger fossil bones.

DO NAUTILOIDS IN THE REDWALL LIMESTONE RECORD A "FLOOD OF CATASTROPHIC PROPORTIONS?"

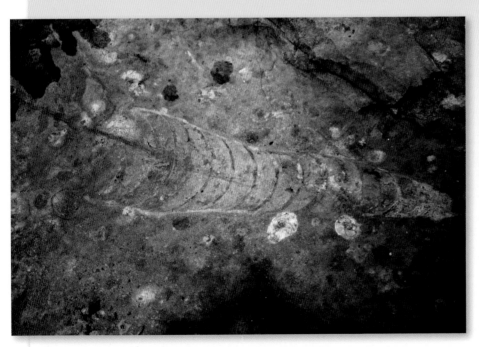

Figure 13-19. Orthocone nautiloid and disarticulated crinoid remains, Thunder Springs Member, Redwall Limestone. *Photo by Howard Lee.*

The Thunder Springs Member of the Redwall Limestone has a roughly 10-foot-thick layer containing a huge number of long, slender nautiloid shells, positioned as would be expected if they had been deposited in a strong current (Fig 13-19). Flood geologists frequently cite the orientation of these shells as evidence of a flood of catastrophic proportions. But virtually all other fossils in the canyon layers are *not* oriented. Using flood geology logic, this must be powerful evidence that a flood of catastrophic proportions *did not* deposit any other layers – only the nautiloid bed.

And recall the petrified logs and ferns. If sorting is the answer, why are logs and ferns still found in the same deposits?

Communities of Fossils

During the centuries between 1550 and 1850, a great number of natural scientists came to the slow realization that fossils occurring in clusters were samples of discrete ancient living communities or *ecosystems*. Early geologists of the nineteenth century recognized that these biologic communities flourished for a limited period of time and eventually were replaced by later communities with a unique suite of life forms, often as a result of apparent shifts in local or global environmental conditions. The succession of biological communities over time is illustrated in dramatic fashion in the combined Grand Canyon and Grand Staircase sequence.

To sum up: Fossils are not just found in a particular order, but also in orderly patterns. These patterns consist of characteristic groupings of fossils that reflect the type of environment in which the community lived, the ecosystem dynamics, and often the changes that eventually led to replacement of one grouping by another. These patterns, which are repeated across the globe, testify to a long succession of living communities (not dead organisms dumped in some huge catastrophe).

What would be the pattern of fossil organisms emplaced in a single, violent catastrophe? If organisms were ripped out of their life environment, carried vast distances by powerful tsunami-like currents, and dumped into layers of sand and mud that were rapidly being deposited, shouldn't we expect that all kinds and complexities of fossils would be jumbled together? The fossils in the Grand Canyon tell us that they – and the rock they are in – were *not* deposited in a single global flood. This is the same story told in the other chapters throughout this book.

MUTUALLY EXCLUSIVE ARGUMENTS

Young Earth advocates argue that broken fossils are evidence of the violence of the flood. Yet they also argue that exquisite preservation of delicate parts of other fossils is evidence of rapid burial by a global flood (Fig 13-20). How are fragmented fossils AND delicately preserved fossils both evidence of a violent global deluge?

Figure 13-20. Fossil of the brachiopod *Peniculauris bassi* from the Kaibab Formation, with some of its delicate spines intact. *Photo by David Elliott.*

Petrified logs at Petrified Forest National Park. *Courtesy of Petrified Forest National Park.*

CHAPTER 14

TINY PLANTS – BIG IMPACT
POLLEN, SPORES, AND PLANT FOSSILS

by **Joel Duff**

Principle of Floral Succession

THE PRINCIPLE OF FAUNAL SUCCESSION technically refers to successive changes in the fossils of animal life (faunal = animals) from one Earth layer to the next. Very similar kinds of changes are observed with fossil plants, so we can also speak of the *principle of floral succession* (floral = plants). As is the case with animal fossils, we find with plants that certain groups of plant fossils commonly are found together in rock layers of the same age, with a distinctly different group of fossil species in the underlying and overlying rock layers. We also find that the complexity or diversity of plant fossils generally increases in progressively younger layers. For plants, the evidence of succession was initially based on studies of macroscopic fossils (visible without magnification) like petrified wood and leaf impressions, but the same succession is found in microscopic fossils of pollen and spores. Pollen and spores play a particularly important role in our understanding of the history of the Grand Canyon and the Grand Staircase. Their value is related to their size, their abundance and distribution, and our ability to identify the type of

plants they came from. The very small size allows them to be distributed broadly by even gentle winds or water currents, and they are often preserved with little damage or alteration. When ease of distribution is coupled with enormous production rates, the result is an abundance of fossil pollen and spores, often present in ancient deposits even if no macroscopic plant tissue was deposited or preserved.

These fossils offer a great way to test the flood and conventional models. In a violent global flood, we should expect a thorough mixing of all types of these tiny particles in all the different flood layers, and we should expect many rock layers to have mismatches between the type of pollen or spores present and the macroscopic fossil plants. In the conventional geology model, where deposits were forming under different environmental conditions over long periods of time, we should expect to see unique types of pollen and spore separated into distinct layers that fit with the types of macroscopic plant fossils found in the same layers or in layers of equivalent age. So what do we actually find? We'll look first at the type and distribution of macroscopic plant fossils found in the Earth's layers, and then focus on the pollen and spores.

Era	Geologic Period	Examples from the Floral Fossil Record in Grand Canyon and Grand Staircase Rocks	
Ceno-zoic	Paleogene and Neogene	Flowering plants and associated pollen become common in the fossil record.	
Mesozoic	Cretaceous	Flowering plant groups appear early in the period. Leaf fossils show a relatively diverse assemblage dominated by sycamore-like plants (Dakota Fm).	
	Jurassic	Conifers remain prominent (Morrison Fm).	
	Triassic	200 species of plants identified including ferns, conifers, and cycads with scars of insect boring (Chinle Fm).	
Paleozoic	Permian	Fragmentary remains of seed ferns (Pteridospermophytes), horsetails (*Calamites*), ginkgos, and other conifers are present (Hermit Fm).	
	Mississippian and Pennsylvanian	Lycopod trees, such as *Lepidodendron*, and ferns dominate the land as witnessed by leaf and spore fossils. *Calamites*, a primitive horsetail, is also present (Surprise Canyon Fm).	
	Devonian	Marine only; no terrestrial plant fossils found.	
	Cambrian	No terrestrial plants known from these formations.	
Proterozoic	Late Proterozoic	Marine algae only; no terrestrial flora known.	
	Early Proterozoic	No fossils in crystalline basement rock.	

Figure 14-1. Summary of the floral fossil record in Grand Canyon/Grand Staircase and global layers.

Flowering plants dominate and diversify. Sunflower and bean families first appear, along with increasing numbers of orchids and grasses.

orchids

sunflowers

first grasses

Most major groups of Angiosperms (flowering plants) appear, becoming the common vegetation type. Grasses only appear very late in the Cretaceous, as do members of the orchid family. Notably absent are any members of the sunflower (Asteraceae) and bean (Fabaceae) families. Lycopods and ferns are represented almost solely by small lower-story forest vegetation rather than the dominant type as seen in the Paleozoic era.

Sycamore (hardwood trees)

Conifers remain dominant forest type, with additional families appearing, such as cypress and yew. Ginkgoes common. The first flowering plants and their pollen appear in the late Jurassic, though are only a very small part of the terrestrial flora.

first flowers

Conifers become prominent, with large lycopods taking a lesser role in plant communities. The conifer-like Cordaites disappear. Tree-ferns common.

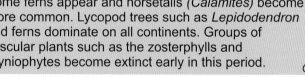
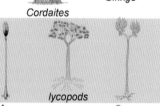

tree ferns

conifer forests

Pine-like conifers called Cordaites become dominant forest type late in the Permian, with cycads and ginkgoes also appearing. *Calamites* disappear in the early Permian. Tree ferns remain common, but some lycopods, including Lepidodendron, go extinct in the early Permian.

Cordaites

Ginkgo

Some ferns appear and horsetails *(Calamites)* become more common. Lycopod trees such as *Lepidodendron* and ferns dominate on all continents. Groups of vascular plants such as the zosterphylls and rhyniophytes become extinct early in this period.

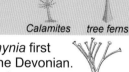

Calamites   *tree ferns*

lycopods

Simple vascular plants such as *Cooksonia* and *Rhynia* first present, but most kinds are extinct by the end of the Devonian. Bryophytes (mosses) are present and lycopods and some horsetails *(Calamites)* first appear but are not common.

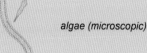

small plants (< 1 foot)

Only algae and fungi on land. (Most major invertebrate phyla appear in the marine fossil record).

algae (microscopic)

Marine algae only – no terrestrial flora known.

blue-green algae (microscopic)

stromatolite

Figure 14-2. Diorama of life in Precambrian time, showing stromatolite mounds along an ocean shoreline. There is no fossil record of land plants existing at this time. Smithsonian Institution diorama, Washington, D.C. *Photo by Tim Helble.*

Plants through Time

The table on the previous two pages shows the different plant types that have existed in the Grand Canyon/Grand Staircase region and worldwide through Earth's history (Fig 14-1). Note that the complexity and diversity of plant life increases over time as one goes from older rock up to younger rock in the sedimentary sequence. As with faunal succession, this is not a hypothetical listing; it is the fossil record of what is actually observed in the Grand Canyon/Grand Staircase and global sedimentary layers.

Prior to the Cambrian Period, the only photosynthetic life forms on Earth were marine algae, and

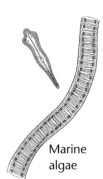

Marine algae

no plants had as yet colonized the land (Fig 14-2). How do we know this? Because no rocks of Precambrian age, in the Grand Canyon or anywhere in the world, have any type of plant fossils in them — just different kinds of algae. This conclusion is not a result of any precon-

ceived assumptions about increasing complexity over time; paleontologists infer the history of plant (or animal) life on Earth from fossils that are actually contained in rocks worldwide.

To find multicellular plants, we have to move up into the overlying Paleozoic layers, but here also we find a distinct order. First, in the Cambrian, there were only algae and fungi on land; next came simple, now-extinct, vascular plants (plants with specialized tissues for transporting water and minerals); then lycopods (seed ferns) and other ferns appeared and dominated until the Permian. Eventually conifers (such as pine trees) dominated until the Cretaceous when flowering plants experienced rapid diversification (Fig 14-3).

Significantly, this pattern is repeated in rocks of similar age all over the world. The most logical explanation, independent of any assumptions about the history of life, is that the oldest rocks formed prior to the appearance of life forms on Earth, and that each successive rock layer preserves a record of the life forms present during that interval of time. How else can one account for the complete absence of

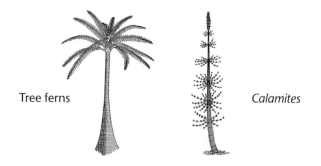

Tree ferns   Calamites

Figure 14-3. Typical swampland forest in the Permian, based on an actual excavation of about an acre of fossilized forest in the Wuda region, China. Dark green trees on left are primitive conifers; tallest "trees" on right are extinct lycopods that reached 50+ feet. Closely related plants have been found in the Hermit Formation. None of these plants exist today. *Illustration courtesy of National Academy of Sciences.*

Figure 14-4. Seed fern, *Callipteris conferta,* from the Hermit Formation. *Photo courtesy of Grand Canyon National Park GRCA #3086.*

fossils in the lowest sedimentary layers worldwide, the absence of multicellular plants in thousands of feet of Supergroup rock in the Grand Canyon, and the complete absence of flowering plant fossils in all of the Grand Canyon layers?

What Are Pollen and Spores?

Pollen and spores are microscopic reproductive products of plants that, due to their environmentally resistant outer walls, are among the most easily fossilized living structures. Spores are typically, but not always, larger than pollen grains and can be thought of as the "seeds" of ferns, lycopods, and mosses. Pollen is the male sex cell of flowering plants and gymnosperms (cone-bearing plants such as pine and spruce trees).

The extraction of pollen and spores from sedimentary rock is not easy and frequently requires treatment with very strong chemical agents that dissolve inorganic material and leave the organic walls of the spores and pollen intact. Following preparation of samples, the fraction of material containing organic remains can be microscopically scanned to identify spores or pollen. The external appearance of these reproductive products is highly variable for land plants, allowing major plant groups to be distinguished from one another. As such, their presence in rocks is a strong indicator of the types of plants that were growing in a region when the sediments were deposited.

Pollen and spores demonstrate the same principle of floral succession in the Grand Canyon and Grand Staircase that the macrofossils do (Fig 14-1, pages 146-147). Cambrian and Devonian rocks, represented by strata from the Tapeats Sandstone up to the Temple Butte Formation, contain only simple trilete (three-lobed) spores, which were generated from now-extinct simple spore-bearing plants and algae (Fig 14-3). Unmistakable spores from lycopods and ferns are plentiful in rocks above the Temple Butte Formation. For example, such fossilized spores are found in abundance in the Surprise Canyon Formation and in higher layers. These spores are associated with the plant remains of ferns, horsetails *(Calamites)* and giant lycophyte trees *(Lepidodendron)*.

Lepidodendron

Cordaites

Fossil pollen from conifer-like (non-flowering) trees (Cordaites) is first encountered in the upper rock formations of the canyon, including the Supai and Coconino Sandstones. Conifers are the dominant plants in Triassic and Jurassic rocks of the Grand Staircase (Fig 14-1, pages 146-147), and their pollen is readily distinguishable from flowering-plant pollen.

Significantly, the absence of any macroscopic fossils of flowering plants in Grand Canyon rock layers is matched by a complete absence of fossil flowering-plant pollen – which fits well with the conventional geology model. Remnants of old flowering plants in the Grand Canyon are found only in the sediments of modern caves, deposited after the major incision of the Grand Canyon (i.e., long after the rock units were laid down; see Chapter 18). The lack of fossilized flowering-plant pollen in the Grand Canyon's sedimentary rocks is exactly what the conventional geology model would expect, because flowering plants (and their pollen) all over the Earth (not just in the Grand Canyon) are found only in rocks of Cretaceous age or younger (about 140 million years ago to the pres-

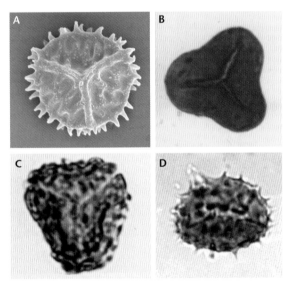

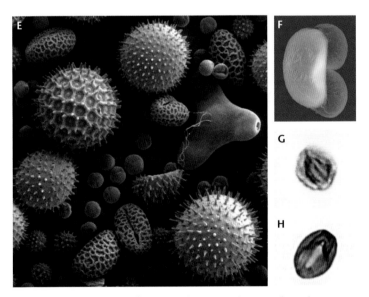

SPORES: (A) Scanning electron microscope image of a modern spore of a simple moss-like plant, showing a typical three-lobed (trilete) shape. (B-D) Light-microscope images of fossil spores extracted from the Grand Staircase: Cretaceous Dakota Formation, Canyonlands area, southern Utah.

POLLEN: (E) Scanning electron microscope image of modern pollen grains from common flowering plants such as sunflower, lily, morning glory (largest pollen grain), and beans (smallest). (F) Light-microscope image of a typical pollen grain of a pine (conifer) tree with diagnostic "wings." (G-H) Pollen fossils extracted from the Grand Staircase: Cretaceous Dakota Formation, Canyonlands area.

Figure 14-5. Modern and ancient spores and pollen, shown at approximately the same magnification. The largest (A) is less than one-tenth of a millimeter in diameter (0.004 inch). *Image credits: (A) Christine Cargill © Centre for Australian National Biodiversity Research; (B-D, G-H) Barbara Anne am Ende; (E) Dartmouth Electron Microscope Facility, Dartmouth College, (F) Joel Duff.*

Figure 14-6. A common modern flower found in the Grand Canyon – Indian Paintbrush. *Photo by Mike Koopsen.*

ent). Not surprisingly, fossilized flowering-plant pollen is found in Cretaceous rocks of the Grand Staircase series. And remember, this ordering of fossils is not assumed, it is the actual order found in the continuous sequence of rocks in the Grand Canyon and Grand Staircase!

As with the distribution of animal fossils, the plant fossils found in rocks around the globe present a serious challenge to flood geology models. While flood geologists have paid much attention to explaining plant macrofossil distribution by various differential-sorting mechanisms

Sunflower

(see Chapter 13), they have made few attempts to explain the distribution of microfossils (spores and pollen) in the fossil record. As described earlier in the chapter, the flood geology model would predict that microscopic remains – such as pollen and spores – of any flowering plants that were present prior to the global flood and distributed worldwide by that flood, should be distributed throughout the geologic column. Yet such remains are conspicuously absent from all rocks older than Cretaceous age. Said another way, in spite of hundreds of square miles of exposed sur-

faces throughout the Grand Canyon, not a single unambiguous pollen grain from a flowering plant has ever been discovered in any unaltered rock layer below the rim. This alone should be enough to categorically reject a global flood explanation for the Grand Canyon layers.

IS POLLEN FROM FLOWERING PLANTS FOUND IN ANY GRAND CANYON ROCKS?

Some flood geologists claim to have found pollen from flowering plants in a sample from the Hakatai Shale of the Precambrian Unkar Group. If true, this pollen would present a challenge to the modern geologic understanding that fossils of flowering plants are present only in Cretaceous-age and younger rock. Other flood geologists have been critical of these claims, however, recognizing them as likely resulting from contamination of the samples with modern pollen. Indeed, the pollen reported is very similar to the types of pollen produced by plants found *locally* in the Grand Canyon *today*. More importantly, if seed-bearing plants were common before the flood (Genesis 1:11), shouldn't this pollen be abundant in most layers?

Tracks in a fallen section of Coconino Sandstone. *Photo by Tom Bean.*

CHAPTER 15

TRACE FOSSILS
FOOTPRINTS AND IMPRINTS OF PAST LIFE

by **David Elliott**

Now WE COME TO A THIRD category of fossils found in Grand Canyon and Grand Staircase rocks. So far we have described *macrofossil* evidence for past animal and plant life, such as bones and petrified wood easily seen with the naked eye, and *microfossils*, such as spores and pollen that must be viewed under the microscope. The third category consists of the traces made by the activities of animals while they were still alive.

What Are Trace Fossils?

Trace fossils are tracks, trails, burrows, borings, and other structures made by ancient organisms that are preserved in the fossil record. (Trace fossils were briefly introduced in Chapter 6 as a type of sedimentary structure.) These features generally formed in soft sediment, but they may also have formed in wood or rock, such as borings. Trace fossils differ from body fossils (fossilized body parts of organisms) in that they

are records of an organism's *behavior* rather than being part of the organism itself.

Prior to fossilization, traces, such as footprints, are easily destroyed if the sediment is reworked or moved. This differs from skeletal or shell material, which can be transported from the environment in which the

Figure 15-1. The undersurface of a bed from the Bright Angel Shale, Grand Canyon, showing feeding traces made by trilobites and indicating a shallow marine environment. *Photo by David Elliott.*

animal lived to another one where it is preserved. For example, a land animal that drowns in a river might be carried out to sea and preserved in marine sediments. Animal traces cannot be transported, and are preserved where they actually formed. Trace fossils thus provide direct evidence of the conditions at that place and time, and of the organism's response to those conditions. For this reason trace fossils are widely used in interpreting past environments.

Trace Fossils in the Grand Canyon

Trace fossils are found throughout the Grand Canyon sequence. Although this chapter particularly details what is known about the Coconino Sandstone, trace fossils are also abundant in the Bright Angel Shale, the Supai Group, and the Hermit Formation. In the Bright Angel Shale, which was deposited in a variety of shallow marine environments, traces are common and include a diverse array of tracks, trails, and burrows. Vertical and U-shaped burrows indicate activity by suspension-feeding organisms (such as marine worms) and are similar to burrows found in modern near-shore environments. Most widespread in the Bright Angel Shale are crawling, feeding, and resting traces formed by trilobites as they moved across muddy sediments (Fig 15-1). All of these traces are typical of today's shallow-marine environments in which animals colonize stabilized surfaces; they could not have been made under conditions in which there was a high rate of sedimentation (such as in a global flood).

The Coconino Sandstone

The Coconino Sandstone is a thick sandstone sequence that is well exposed in the Grand Canyon and many other parts of the Colorado Plateau south of the Arizona/Utah state line. The Coconino Sandstone is evidence of an enormous desert sand sea (called an *erg*) that covered an extensive area of what is now the western United States during the Permian Period, roughly 275 million years ago.

Why do geologists consider these deposits to represent desert deposits and not aquatic (water) deposits? As was described in Chapter 6, compelling evidence

Figure 15-2. Two sets of *Chelichnus gigas* tracks in the Coconino Sandstone, eastern Grand Canyon. *Photo by Scott Thybony.*

exists in the form of features within the sandstone that can be compared to similar features that form in desert deposits today. This evidence can be divided into sedimentary features, such as cross-beds, that were created as the sand was deposited, and trace fossils that were formed by the organisms that lived in the ancient desert environment.

Although no actual remains of organisms have been found in the Coconino Sandstone, the traces of animals and their behavior have been preserved as footprints and burrows in the sand (Figs 15-2, 15-3, 15-4). These traces range from large to small vertebrate tracks and also include tracks and burrows very similar to those left by spiders, scorpions, millipedes, and other arthropods in modern desert environments. The Coconino Sandstone contains no evidence of aquatic organisms of any kind that might support an argument for deposition in a deepwater, flood environment, such as has been proposed by flood geologists.

The lack of preserved bones of vertebrate animals is typical for a desert environment. Soft tissues are rapidly scavenged, and bones lying on the surface are exposed to intense solar radiation and extreme drying conditions. In addition, the large daily temperature changes in a desert result in heating and cooling on a regular basis. Together, these effects cause bones to crack and flake, and quickly crumble and vanish. Therefore it is not surprising that no bones of the animals that lived in the Coconino sand-sea desert have been found.

Vertebrate Tracks

The vertebrate tracks found in the Coconino Sandstone consist of sets of prints varying in length from half an inch to four and a half inches. Prints often occur in two rows, which is typical of four-legged animals.

An interesting feature of many Coconino tracks is a pad of sand pushed by the foot at the back of each print, indicating that the animal was moving up the

A. Tracks of *Chelichnus bucklandi*, a small vertebrate, from Ash Fork, Northern Arizona.

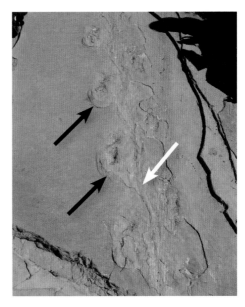

B. Tracks from *Chelichnus gigas*, a large vertebrate, from the Hermit Trail, Grand Canyon.

C. *Chelichnus gigas* tracks, from the Boucher Trail, Grand Canyon.

D. *Chelichnus gigas* tracks, from the Boucher Trail, Grand Canyon. Claw impressions are distinct and smaller millipede trails are also visible moving in the same direction as the vertebrate tracks.

Figure 15-3. Tracks found in the Coconino Sandstone, all made by vertebrates moving up dune faces. Black arrows show sand pushed back behind prints, and white arrows show sinuous tail drag marks. *Photos by David Elliott.*

A. Spider track almost 10 inches across, from Boucher Trail, Grand Canyon.

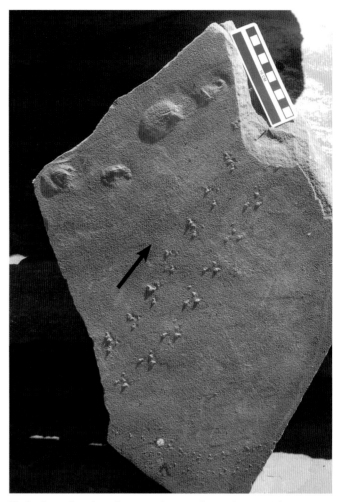

B. Surface showing three tracks: reptile track at top, large spider track in the center, and smaller spider track along the bottom, from Ash Fork, Northern Arizona.

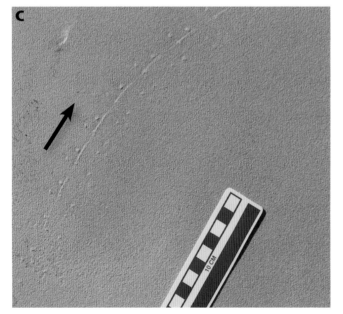

C. Scorpion track showing tail drag, from Ash Fork, Northern Arizona.

D. Isopod (pill bug/roly poly) and millipede tracks, from Boucher Trail, Grand Canyon.

Figure 15-4. Invertebrate tracks from the Coconino Sandstone. Arrows show direction of movement. *Photos by David Elliott.*

sloping face of a sand dune (Fig 15-3B, D). In some of the largest sets of footprints, a sinuous groove between the rows of prints indicates where the animal's tail dragged (Figs 15-3B and C). The footprint area and depth of the tracks are proportional to the weight of the animal; tracks in the Coconino indicate that the largest animals weighed between 20 and 30 pounds. The larger vertebrates were probably active predators, because large tracks have been found that follow or cut into smaller tracks. In these cases, widely separated footprints indicate that the predator was running rapidly. It used to be thought that a large number of species were responsible for the various tracks, but further investigation has recently resulted in the number being reduced to only three species, all belonging to the genus *Chelichnus*. The low number of species is also consistent with the low number of vertebrates expected for a dune-field environment.

Although there are no skeletal fossils for the vertebrates that made the tracks in the Coconino Sandstone, a deposit of the same age in Texas contains similar footprints associated with the bones of early synapsids (creatures with both reptile-like and mammal-like skeletal features). This suggests that the Coconino vertebrates may also have been early synapsids.

Arthropod Tracks

Trace fossils left by invertebrate organisms, such as insects, spiders, millipedes, and scorpions, include tracks made as the animals traveled over the sand's surface, and burrows created as they buried themselves in the sand. Some distinctive tracks consist of alternating sets of three or four impressions in a double row, sometimes with an impression left by a dragging tail. Experiments with living organisms have shown that these fossil tracks are most similar to modern tracks made by spiders (Tarantula, Trapdoor, and Wolf spiders) and scorpions. In some cases these tracks are as wide as one foot across, which indicates the presence of very large spiders in the Coconino desert (Figs 15-4A, B). Studies on modern spiders and scorpions show that variations in temperature and surface conditions can result in the same animal leaving a variety of traces, alternately leaving prints from two, three, or four legs on each side. In the case of

scorpions, sometimes there is no tail drag, but in other cases there is an intermittent one or a continuous one. These variations are also present in tracks from the Coconino Sandstone (Fig 15-4C). Other trails in the Coconino are similar to those produced by modern millipedes and isopods (pill bugs/roly polies) that are common today in desert environments (Fig 15-4D).

How Trace Fossils Were Preserved

The method of preservation for delicate prints such as the ones shown in these photos has been well investigated. Experimentation with spiders, scorpions, and small lizards has shown that tracks similar to those in the Coconino Sandstone can

Figure 15-5. Arthropod and small reptile tracks in Namib Desert sands. *Photo by David Elliott.*

easily be formed in dry or moist sand, but that it is difficult or impossible (particularly for small animals) to form these kind of tracks in sand that is under water. Preservation of tracks after they have been made requires either hardening of the imprinted surface by evaporation of moisture that leaves behind mineral precipitates to hold grains together, or the formation of biofilms (microbes that attach to the sand grains and cause them to adhere to each other).

Where would water come from in a desert? Modern deserts typically receive less than six inches of mean annual rainfall. Raindrop impressions can be seen in the Coconino Sandstone (see Fig 6-5, page 69), but direct rain would disrupt surface traces rather than preserve them.

A good example of how tracks can be preserved in a sand-dune environment is found in the Namib Desert in southwest Africa, where mists are generated offshore over the ocean and move inland almost daily (Fig 15-5). These mists form the primary water source for many small desert animals and for consolidation of the dune faces. Since sedimentary rocks indicate there were marine conditions to the west of the Coconino desert in Permian time, mist has been proposed as a water source and mechanism for preserving the tracks of vertebrates and arthropods in the Coconino until they were covered over by dry blowing sand.

Were the Coconino Tracks Made by Amphibians Walking Underwater?

Some flood geologists have claimed that the vertebrate tracks present in the Coconino Sandstone were made by animals, likely amphibians, walking or running underwater in an attempt to escape advancing flood waters. To support this claim, studies were performed with amphibians in tanks of water and sand to demonstrate that tracks can be made underwater. Aside from ignoring all the physical evidence for a desert environment in these sands, the water-tank experiments fail to account for the Coconino tracks on several counts. They do not replicate the presence of running and galloping gaits that are possible only on dry land,

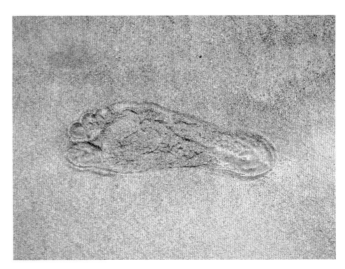

A. Footprint made in sand along the Colorado River in the Grand Canyon.

B. Gentle wave washes over footprint.

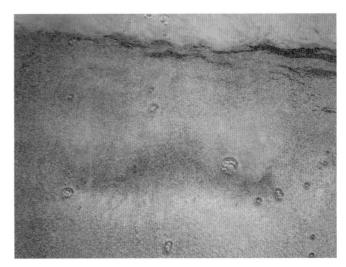

C. Wave washes out, almost completely obliterating the footprint.

Figure 15-6. Fate of a footprint passed over by a gentle wave of water. *Photos by Tim Helble.*

and the experiments, ironically, were performed in a *gentle* current of 0.2 mi/hr (3 in/sec) – hardly the raging torrents required by the flood geology model. Perhaps this is because the experimenters knew that even mild flows of one foot per second quickly wipe out footprints. Just consider the fate of a footprint on the beach when the gentlest of waves passes over it (Fig 15-6).

What Can Be Concluded from the Evidence?

Taken as a whole, the sedimentary and trace fossil evidence clearly and unambiguously points toward a desert environment for the Coconino Sandstone (Fig 15-7). Although flood geologists claim that this sandstone was deposited in rapidly flowing water that was hundreds of feet deep, the presence of raindrop impressions, desiccation cracks, and well-sorted frosted sand grains, along with the angle of slope in cross-beds, the absence or scarcity of body fossils, and the preservation of footprints, are all clear indicators of a sand desert. Asserting an aquatic environment for the Coconino Sandstone requires ignoring nearly all the geological and paleontological evidence. Truly, as one flood geologist has said, "I wouldn't have seen it if I hadn't believed it."

Figure 15-7. A reconstruction of the environment in which animals lived in the Permian Coconino sand desert.
Art by Casey Brose.

Grand Falls on the Little Colorado River. *Photo by Tim Helble.*

PART 4

Carving of the Canyon

We now come to a time after all the rock layers of the Grand Canyon and Grand Staircase had been deposited and erosion began to take over, first stripping the thick Grand Staircase layers away from the Grand Canyon area, and then carving the deep canyon we see today. This general sequence of events is agreed upon by modern and flood geologists. Flood geologists contend that the once-continuous Grand Staircase layers were stripped away from the Grand Canyon region as the global flood waters "decreased off the earth continually" (Genesis 8:3), with the canyon forming rapidly in still-soft sediments, sometime after the flood, by catastrophic post-flood processes. The conventional geology understanding is that the canyon formed in hard rock over millions of years.

This section is divided into three parts. The first chapter (Chapter 16) is an assessment of the most common flood geology arguments for rapid carving of the canyon. The second chapter (Chapter 17) describes the conventional geology understanding of how the canyon was carved, with a primary focus on the age of the canyon. The third chapter (Chapter 18) focuses on what we can tell about the canyon's history since the time of carving, based on more recent material found filling caves along the canyon walls.

A muddy torrent spills over the Redwall Limestone in Marble Canyon. *Photo by Rob Elliott.*

CHAPTER 16

CARVING OF THE GRAND CANYON

A LOT OF TIME AND A LITTLE WATER, A LOT OF WATER AND A LITTLE TIME (OR SOMETHING ELSE?)

by **Tim Helble** and **Carol Hill**

How was the Grand Canyon carved? When Young Earth advocates address this question, they typically consider two options. Was it a lot of time and a little water, or a lot of water and a little time? The impression given is that these are the only two possibilities, with modern geologists favoring the first option – that the canyon formed over eons of time – and flood geologists favoring the second option – that the canyon formed catastrophically a short time after a global flood. As we will see later in this chapter, an important third option is left out.

In evaluating these options, we will consider three common arguments made by flood geologists to support a rapid and recent carving of the canyon. The first is the "breached-dam hypothesis," which is

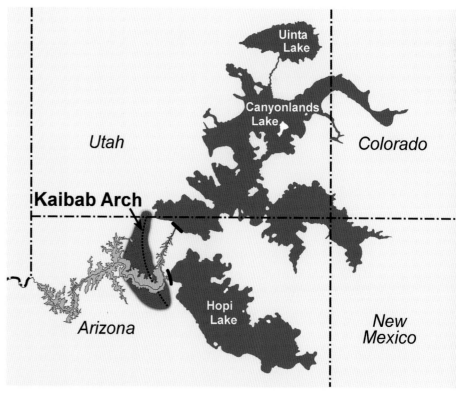

Figure 16-1. Proposed lakes that emptied catastrophically to carve the Grand Canyon, according to flood geologists. *Figure based on W. Brown:* IN THE BEGINNING: COMPELLING EVIDENCE FOR CREATION AND THE FLOOD; *S. Austin:* GRAND CANYON: MONUMENT TO CATASTROPHE; *and Answers in Genesis's Creation Museum in Kentucky.*

used to argue that the Grand Canyon was carved by the catastrophic failure of dams associated with giant post-flood lakes. The second is the "where did all the sediment go?" argument, which is used to infer that geologists are stumped about "missing" sediment that was excavated from the Grand Canyon. The third is the "rapid carving of small canyons" argument, which is used to equate the rapid carving of small canyon systems around the world with a rapid carving of the mile-deep Grand Canyon.

1. The Breached-Dam Hypothesis

Perhaps the most intriguing question regarding carving of the Grand Canyon is how the Colorado River managed to cut a channel through the Kaibab Arch (see Fig 3-3 map, page 33). The crust at this location bows upward, creating an arch that rises 3,000 feet above the land on either side (see Fig 12-12, page 124). How did a canyon form that allowed the Colorado River to flow *through* the arch instead of *around* it?

Flood geologists maintain that a number of lakes were left on the landscape after the end of Noah's flood. The most prominent flood geologists propose that the Kaibab Arch and other features were rapidly uplifted near the end of the flood, creating natural dams that formed large lakes upstream from the present-day Grand Canyon. The proposed ancient lakes are identified as the Canyonlands and Uinta Lakes to the northeast of the Grand Canyon and Hopi Lake to the southeast (Fig 16-1). Flood geologists have estimated that together, these lakes once contained over 3,000 cubic miles of water – almost three times the volume of Lake Michigan. An unspecified number of years later, the dams failed catastrophically and lake water burst out, carving the Grand Canyon in a matter of days. Flood geologist Henry Morris summarized the "breached-dam hypothesis" this way: "...a great dammed-up lake full of water from the Flood suddenly broke and a mighty hydraulic monster roared toward the sea, digging deeply into the path it had chosen along the way."

Flood geologists aren't the only ones who have argued for a breached dam at some point in the history of the Grand Canyon. Sudden emptying of a lake – although at a much smaller scale than that proposed by flood geologists – has been endorsed by some non-flood geologists, who refer to it as the "spillover" model. This idea became popular among many geologists at the symposium on the Origin of the Colorado River, held at the Grand Canyon in 2000. The ancient lake in question is Lake Bidahochi (sometimes referred to as Hopi Lake), located in the region where the Little Colorado River exists today. When the spillover model came out, it was immediately seized upon by flood geologists because it appeared to confirm their breached-dam hypothesis. Though seriously considered for a time by conventional geologists, the spillover model and the breached-dam hypothesis both are now rejected by many Grand Canyon geologists because the bulk of the evidence does not support either theory.

Figure 16-2. Hiker standing on an igneous dike (Big Hopi Butte dike) that cuts across the lower member of the Bidahochi Formation, about 10 miles south of Dilkon, Arizona. *Photo by Michael Ort.*

Figure 16-3. Bidahochi Formation. Note the lake layers demarked by the color banding in the formation. *Photo by Wayne Ranney.*

Is there any evidence for the proposed lakes?

A serious problem with the breached-dam hypothesis is the lack of evidence that these lakes were as extensive as claimed, or if they even existed at all. Ancient lakes are readily identifiable by the characteristic deposits that form when sediment carried by rivers flows into a freshwater reservoir, slows, and settles onto the lake bottom. Do we find vast lake deposits at the proposed locations? Canyonlands Lake, which flood geologists say extended all the way into Colorado, is purely hypothetical and has no physical evidence to support it. Unita Lake did exist, though a connection between Unita Lake and Canyonlands Lake is even more hypothetical. Justification appears solely based on needing extra water to accomplish carving the Grand Canyon after a dam failure. There is also genuine evidence for Hopi Lake in the form of lake deposits of the Bidahochi Formation (Figs 16-2, 16-3). The Bidahochi Lake deposits are known to range from 6 to 16 million years in age because a number of radiometrically dated volcanic dikes and ash flows intrude and interbed with these lake deposits.

Recent scientific studies of Bidahochi sediments show that the lake pictured as Hopi Lake by flood geologists was probably never one big lake. Rather, it consisted of a series of lakes or playas (lakes that seasonally dried up) that were never able to spill over or breach a dam. In fact, evidence for a spillover or failure point for this proposed lake has yet to be found.

A 600-FOOT CLIFF OF LOOSE LIME?

Erosion down into an un-cemented deposit of lime or shells can form a short vertical face, but what happens as downcutting exposes a taller and taller cliff face? Few cliffs survive exposures of more than a few tens of feet before collapsing to form a gentler slope. A 600-foot limestone cliff face, like the Redwall Limestone in the Grand Canyon, is clear evidence that the cliff was already hard rock when carved.

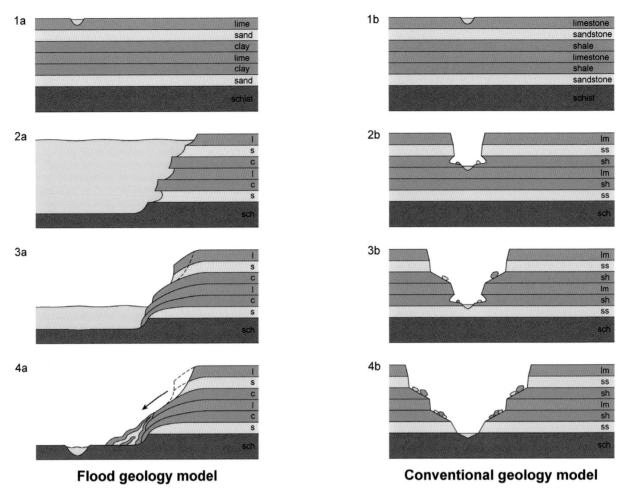

Flood geology model **Conventional geology model**

Figure 16-4. Predicted behavior and appearance of the Grand Canyon walls if (left side) floodwaters from a failed giant dam suddenly cut into soft sediment, versus (right side) normal spring floods gradually cut into rock. Sediment or rock type is identified in step 1 for each model, with symbols for the same layers in steps 2-4. (See text for discussion.)

Were the Grand Canyon's layers still soft when the dams were breached?

Most flood geologists argue that the canyon was carved before the layers of sediment had turned into rock. Rushing waters are called upon to explain the main canyon, with side canyons forming as soft sediments collapsed into the main canyon and then were washed away. Observing eroding layers around the world, we have a good idea of what should happen if cliffs hundreds of feet thick form in rock versus in sediment. The figure above illustrates what we should expect to see for the two competing scenarios: on the left, a single raging flood carving down into layers of soft sediment, and on the right, multiple cycles of seasonal floodwaters gradually incising downward through layers of hard rock (Fig

16-4). In the first scenario, where the layers are still soft, sand and lime will wash away much more easily than compacted clay, so the massive burst of water (step 2a) should leave clay layers sticking out a short distance into the channel (i.e., no flat shale "benches"). As the water recedes (step 3a), the weight of overlying layers should cause the soft sediments, especially clay and lime, to extrude like putty into the channel, resulting in thinning and sloping downward near the edge. Finally, slumping should leave piles of mixed sediments at the base of the exposed embankments (step 4a).

For the conventional geology model, rivers should carve downward into the already-hardened rock, leaving vertical walls behind. Erosion of weaker rock layers during seasonal flooding should undermine the cliff faces (step 2b) and result in collapse and widening of the canyon (step 3b). The layers should not thin or slope downward near the cliff edges, because the layers are solid rock. Shale, being much softer than sandstone or limestone, erodes faster, resulting in the collapse of overlying harder layers and producing an ever-widening canyon with shale benches and sandstone and limestone cliffs (step 4b). Debris from higher layers will be common on the benches but less common at river level, where seasonal floods wash it away.

So what is actually observed? None of the expected features for the flood geology model are observed. All the expected features from the conventional geology model are observed.

Could breached dams provide enough water to do the job?

Does the breached-dam hypothesis provide enough water to have scoured 1,000 cubic miles of sediment from the 278-mile-long Grand Canyon? Three thousand cubic miles of water mixed with one thousand cubic miles of sediment would become less like water and more like creamy mud. Even if a gully or small canyon already existed on the west side of the Kaibab Arch before a dam was breached, there would be nothing to keep the huge volume of sediment-choked water from quickly fanning out into multiple channels over thousands of square miles, rather than cutting one deep canyon. (See the Channeled Scablands section later in this chapter.) As the muddy waters slogged westward, huge amounts of deposition would have occurred well before the flow arrived at the current end of the Grand Canyon. And, even if the water could eventually cut through 4,000 feet of soft sediment layers left by Noah's flood and reach the basement rock, by that time most of it would have drained out of the Grand Canyon area. Any remaining water wouldn't have nearly enough "punch" to do the hardest part of the job: cutting through another 1,000 feet of very hard metamorphic and igneous basement rock to form the Grand Canyon's Inner Gorge.

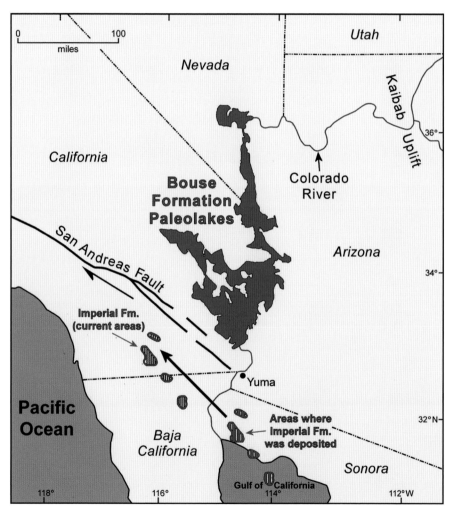

Figure 16-5. Map showing areas where Colorado River sediment has been found below the Grand Canyon. The Bouse Formation was deposited in a series of lakes along the Lower Colorado River Corridor. The Imperial Formation is a marine to deltaic formation. The black arrow shows that the Imperial Formation has been moved northwestward by the San Andreas Fault over the last 6 million years or so. *Figure modified from J. E. Spencer et al., 2010, U.S. Geological Survey, Open File Report 2011-1210.*

2. Where Did Sediment Excavated from the Grand Canyon Go?

In an effort to cast doubt on the conventional geologic understanding of how the canyon was carved, flood geologists ask a

Figure 16-6. Lenore Lake, Washington, in the Channeled Scablands, showing the impact of a megaflood on the landscape. *Photo by Victor Baker.*

puzzling question: Where is all the sediment that supposedly was steadily eroded from the canyon over millions of years? The question is typically followed by the answer that we should expect it to be found somewhere between the canyon and the sea but geologists can't seem to locate it. What makes the question so puzzling is that if that answer were true, the flood geology model would offer no solution. Sediment carved by floodwaters from a breached dam should also be found somewhere "between the canyon and the sea."

In truth, geologists know *exactly* where this eroded sediment is, as has been documented in numerous papers on the Grand Canyon. Flood geologists have either not read the relevant literature, or they simply selectively report results that fit their model. The supposed missing sediment is found within the Bouse Formation southwest of the canyon, in the Imperial Formation in California, and in the Gulf of California (Fig 16-5). While the existence of this sediment might initially be thought to support either

the conventional geology or flood geology models for canyon carving, it actually does little to support a catastrophic explanation. A brief, raging flood should have left deposits of angular rock fragments below the canyon, but such fragments are entirely absent. In contrast, the well-rounded river gravels that are actually observed fit well with a long history of Colorado River flow and normal seasonal flooding.

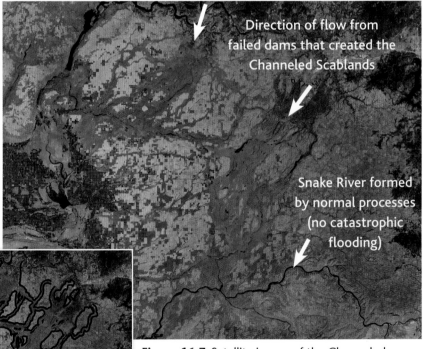

Direction of flow from failed dams that created the Channeled Scablands

Snake River formed by normal processes (no catastrophic flooding)

Figure 16-7. Satellite image of the Channeled Scablands. Inset repeats image with the broad eroded channels outlined. Contrast the wide, shallow channels of the Channeled Scablands with the normal canyon carved by the Snake River. *Copyright 2013 TerraMetrics, Inc., www.terrametrics.com.*

Figure 16-8. Satellite image of the Grand Canyon region. Note how the shape of the Colorado River through Grand Canyon looks more like the Snake River (Fig 16-7, bottom) than the Channeled Scablands. *Copyright 2013 TerraMetrics, Inc., www.terrametrics.com.*

Flood geology locations of failed dams creating the Grand Canyon

3. Rapidly Formed Small Canyons Equated with Rapidly Formed Grand Canyon

To support the idea that the Grand Canyon was carved rapidly, proponents of flood geology often cite other canyons known to have been eroded over short time frames. We will briefly discuss four examples: the Channeled Scablands in eastern Washington, the small canyons below Mt. St. Helens in southwestern Washington, Canyon Lake Gorge in central Texas, and Providence Canyon in western Georgia.

Channeled Scablands

The Channeled Scablands of eastern Washington is a vast area of erosional features carved into a landscape of lava rock (Fig 16-6). The Scablands display channels surrounded by buttes and mesas, with typical relief of about 60 to 300 feet. It is now widely accepted that this unusual landscape was formed by a unique mechanism — a "megaflood" caused by the sudden release of water from a giant glacier-dammed lake known as "Glacial Lake Missoula." Lake Missoula may have contained 500 or more cubic miles of water that abruptly emptied — likely on more than one occasion — as ice dams formed and failed. Peak discharge rates have been estimated to be equiv-

alent to perhaps ten times the combined flow of all rivers in the world.

The existence of such a geologic phenomenon initially appears to lend credence to the flood geology model for the Grand Canyon, so it will likely come as a surprise that the Channeled Scablands provide some of the strongest evidence *against* the flood model. The Channeled Scablands provide an exquisite field site for seeing what is left behind by a megaflood — and that site looks *nothing like* the Grand Canyon! In the satellite image on the facing page (Fig 16-7), we see multiple wide, shallow channels that fan out over a large area — not a deep, narrow, single channel. Additionally, note the total absence of any sharp bends or meander loops in the eroded channels. Megafloods spill over the top of tightly bending channels, carving a new channel that cuts off the bend. Compare this with the winding Colorado River just downstream of the supposed breach in the Kaibab Arch, where raging flood-waters would have been at their most violent (Fig 16-8). The landscape of the Channeled Scablands is a compelling argument against a megaflood origin for the Grand Canyon.

Mount St. Helens

Mount St. Helens is another site favored by flood geologists. The attraction of flood geologists

Figure 16-9. Small canyons and gullies carved into soft ash deposited by the 1980 eruption of Mount St. Helens. *Photos by Tim Helble.*

to Mount St. Helens not only involves the subject of radiometric dating (see Chapter 9), but it also involves the subject of how layers are formed and how quickly canyons can be cut.

When Mount St. Helens erupted on May 18, 1980, nearly a cubic mile of hot pumice rock, ash, and eruption debris catastrophically slid off the north face of the mountain and filled the valley below. About 22 months later, a smaller eruption melted the deep snow in the mountain's new crater and sent a large flash flood of water and mud rushing into the North Fork Toutle River Valley. This flash flood immediately began cutting a new network of channels into the loose, un-cemented ash, rock, and debris left in the valley by the 1980 eruption (Fig 16-9).

These rapidly formed canyons are offered as evidence supporting rapid formation of the Grand Canyon, but little attention is given to anything other

than how fast they formed. The numerous, U-shaped, small canyons and gullies cutting into the loose material below Mount St. Helens do not look anything like the single, massive, V-shaped Grand Canyon. As vertical walls formed in the Mount St. Helens ash, the unsupported ash slumped and dropped piles of loose material down into the stream channels. In great contrast, the Grand Canyon walls remain vertical at heights of hundreds of feet, with no evidence of slumping and no piles of un-cemented and splattered sediment at the base of their cliffs (recall scenarios illustrated in Fig 16-4, page 166). The striking differences between Mount St. Helens and the Grand Canyon provide strong evidence that the Grand Canyon layers were rock when they were carved, not soft deposits like those from Mount St. Helens.

Canyon Lake Gorge

The 2002 Canyon Lake Dam spill-event on the Guadalupe River in central Texas is another event

Figure 16-10. Portion of the gorge cut into the 1.4-mile-long spillway below Canyon Lake, Comal County, Texas. *Photo by Gregg Eckhardt.*

Figure 16-11. Canyon Lake Dam spill event, July, 2002.
Photo courtesy of National Weather Service, Austin/San Antonio Weather Forecast Office.

often cited by flood geologists for rapid carving of a canyon. Yet it, too, provides a compelling argument *against* the flood model for the Grand Canyon. In July of 2002, 25 inches of rain fell in a week's time and sent 25 billion cubic feet of water over the dam's emergency spillway at a peak rate of 67,000 cubic feet per second, scouring out about 16 million cubic feet of rock and soil to form a gorge up to 40 feet deep (Figs 16-10, 16-11).

Why does this event work against the flood model? Recall that the flood model calls for 3,000 cubic miles of water scouring 1,000 cubic miles of sediment and rock from the Grand Canyon, or three times more water than sediment. Canyon Lake Gorge provides an excellent chance to do a reality check on how much water is really needed to accomplish the job. When we do the math, we find the volume of water that poured through Canyon Lake Gorge was *1,500 times more* than the volume of sediment and rock that was scoured! The Canyon Lake Gorge flood offers no support for the flood geology model.

Providence Canyon

Providence Canyon in western Georgia is the result of poor farming practices during the early 1800s (Fig 16-12). If we take a hilly area made up of un-cemented sand, silt, and clay, strip away the tree cover, plow straight downhill, add 50 inches of rain a year, and let soil erosion do its work for 160 years – we can expect the large gullies of Providence Canyon. The pinnacles and narrow walls of Providence Canyon might look similar to

much larger features in the Grand Canyon – such as Zoroaster Temple or The Battleship – but some of those Providence Canyon features have disappeared overnight in heavy Georgia rainstorms because of the softness of the sediment in which they formed. The longevity of the Grand Canyon's buttes and spires speaks to the resistant nature of rock, not to the malleable nature of soft sediment.

Conclusions

So was the Grand Canyon carved in a lot of time by a little water, or by a lot of water in a little time? As we hinted in the title of the chapter, this is really a misleading question. Geologists today do not think it took a little water and a lot of time, but a lot of water *and* a lot of time. If we consider the volume of water that the Colorado River would bring into the Grand Canyon each year if there were no upstream dams (an average of 4.4 cubic miles of water, using recent data), and assume that the rather low average annual precipitation at Phantom Ranch (about 9 inches) fell each year over the entire land area that drains into the Grand Canyon between its upper and lower ends (41,000 square miles), we would estimate that *61 million cubic miles* of water has been eroding the Grand Canyon during the past 6 million years. That's a lot of water *and* a lot of time!

Figure 16-12. Providence Canyon, Georgia. *Photo by Tim Helble.*

Zoroaster Temple from the Colorado River at Pipe Creek, River Mile 89. *Photo by Bronze Black.*

CHAPTER 17

HOW OLD IS THE GRAND CANYON?

by **Gregg Davidson, Carol Hill,** and **Wayne Ranney**

UPON SEEING THE TITLE OF THIS chapter, some readers may wonder why the subject of age is being revisited. Wasn't the question of age already addressed in several preceding chapters? We have indeed already discussed age, but previous material focused on the age of the layers of rock exposed along the canyon walls, not on the canyon itself. In this chapter, we are asking when those layers were *eroded*. How long ago was the canyon carved out, and how long did that process take? This turns out to be a much more difficult question to answer than how old the rock layers are, for what we are now trying to date is not rock – material we can break off and hold in our hands – but a hole in the rock.

By way of analogy, suppose we find an abandoned concrete roadway with no records of its construction, and we want to know more about its history. The chemistry of cement mixtures has changed over the years, so the chemical composition of the roadway may offer clues to when it was made. Saplings growing up through cracks could also be used to set a minimum age for the roadway, because the concrete must predate the oldest tree found breaking through its surface. But now let's say

we want to know when a large pothole formed in this roadway (Fig 17-1). We are now trying to date an event, not the material itself. We can say with confidence that the pothole must have formed after the road-

Figure 17-1. Pothole. *Photo by Miguel Tremblay.*

way was constructed, but how much later, and how long it took to reach its current size, is harder to determine.

So What Is Known about the Age of the Grand Canyon?

Following the pothole analogy, we can start by noting that the canyon must be younger than the rocks it cuts through. The most recently formed rocks are those at the rim – the Kaibab Formation, deposited about 270 million years ago – but even younger layers once lay on top of the Kaibab that are still found today in the Grand Staircase layers (Fig 3-2, pages 32-33). Those younger layers had to have been deposited and then eroded before the canyon could cut through the Kaibab and underlying layers.

Chapter 17 | *How Old Is the Grand Canyon?* | 173

Geologists studying these additional layers and the evidence of regional uplift generally agree that the maximum *possible* age of the canyon is about 80 million years.

And there the agreement ends. There is currently active debate among geologists studying the canyon, with a majority holding to a view that most of the incision occurred within the last 6 million years, and a smaller group arguing that incision of an earlier ("paleo") canyon began long before that. At the heart of the debate is how and when the Colorado River carved its way through the massive Kaibab uplift (or arch), and where water was flowing before the uplift was breached.

The Kaibab uplift is no small feature, rising 3,000 feet above the surrounding region (Fig 17-2). The Kaibab Formation warps upward dramatically from the Marble Platform east of the Grand Canyon, and contributes to the dramatic relief within the canyon. Water does not flow uphill, so how did the Colorado River manage to cut through such a large topographic rise?

There is general consensus among geologists that the Kaibab uplift was initially a drainage divide, with separate river systems developing on either side. The ensuing history is less clear. On the east side of the uplift, the Colorado River currently flows south before making an abrupt turn to the west to pass through the uplift. Tributary canyons on this eastern side oddly point upstream, suggesting to some that the channel here originally developed with drainage to the north. On the west side, drainage to the west would have begun cutting downward, becoming longer in the upstream direction as it eroded back into the uplift (a process called *headward erosion*). Headward erosion on the west side, shown as *Scenario 1* in the figure on the following page, may have eventually worked its way through the uplift to connect with the eastern stream, thus diverting water from the intercepted stream system and redirecting its flow to the route now traveled by the modern Colorado River (Fig 17-3). This pro-

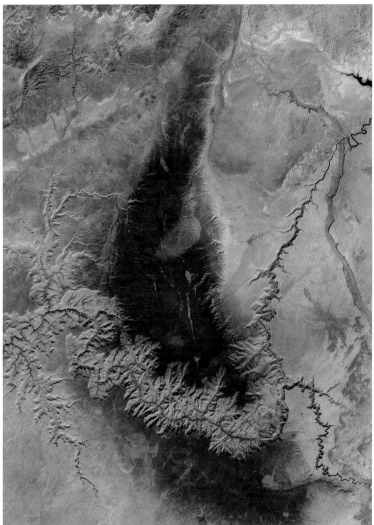

Figure 17-2. Terrain image of the Kaibab Plateau. The darker green represents the Kaibab uplift. *Copyright 2013 TerraMetrics, Inc., www.terrametrics.com*

cess, called *stream piracy*, is a commonly observed phenomenon today.

Now is an exciting time to be a geologist in the Grand Canyon, as recent advances in analytical techniques and an increase in the number of researchers studying this question have resulted in a wealth of new data – and in a lively discussion over the best interpretations to explain the data. Two of the authors/editors of this book, Carol Hill and Wayne Ranney, are in the thick of this debate.

Hill is one of a small number of geologists who have rappelled down or climbed up the vertical canyon walls to reach caves hidden from most people's view (Figs 17-4, 17-6). Ancient mineral formations called *speleothems* (*speleo* = cave, *them* = deposit) in these caves

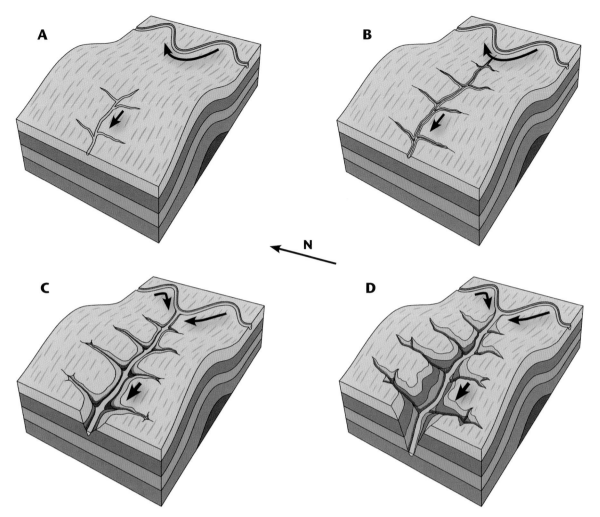

Figure 17-3. Scenario 1: The diagrams represent increasing time from A to D. Streams to the west and east of the Kaibab uplift were initially separate stream systems (A, B), eventually joined by interception of the eastern stream by simple headward erosion of the western stream (C, D). Arrows show direction of river flow.

are adding a whole new chapter to what is known about the canyon's history. Caves, like the canyon, are younger than the rocks they occur in (you can't dissolve a hole in something that isn't already there). By dating certain types of speleothems, Hill and her colleagues have been able to determine when the ancient water table began to drop in elevation as the western canyon was cutting downward.

Hill has also worked on the question of how initially separated streams on the eastern and western sides of the Kaibab uplift could have become connected to form the canyon we see today. She has argued that these streams were likely first con-

Figure 17-4. Author Carol Hill climbing up to a cave entrance in the Grand Canyon. *Photo by Bob Buecher.*

nected by water flowing *under* the Kaibab uplift through fractures and caves, a type of stream piracy called *karst piracy*. Collapse of overlying rock into the caves, shown as *Scenario 2* in the illustration on the following page, eventually opened up a surface passage through which the river then could flow from one side of the uplift to the other (Fig 17-5). This scenario may sound far-fetched to those not familiar with cave processes, but this very phenomenon can be observed occurring today in places like southwestern Germany. In the region of the Swabian Alb, water from the Danube River – flowing eastward toward the Black Sea – suddenly disappears down

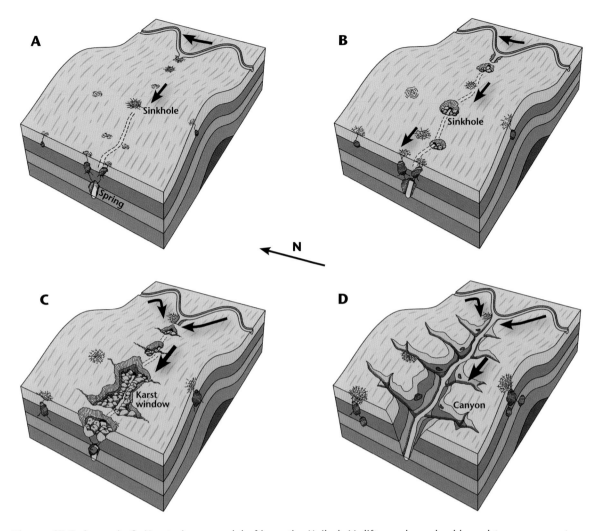

Figure 17-5. Scenario 2: Karst piracy model of how the Kaibab Uplift was breached by subterranean water.

sinkholes, travels over 7 miles through caves, passes under a high ridge, and reappears on the other side to discharge the pirated water into the Rhine River and then north into the North Sea.

Not everyone agrees with the cave model – but that is to be expected when new hypotheses come out. Although there is much yet to be learned about the details of the Grand Canyon's formation, it is clear that the canyon is vastly older than the few thousand years allowed by flood geologists. So stay tuned! As more data are gathered, our picture of the canyon's history is likely to be adjusted and improved.

How Science Works

Young Earth advocates frequently draw attention to disagreements among scientists, such as the opposing theories we have just described, as "proof"

that the conventional scientific view of the Grand Canyon, and indeed of the entire Earth, is based on flimsy evidence and wishful thinking. After all, if the evidence for great age were truly strong, wouldn't all the pieces of the puzzle fit neatly together, and wouldn't everyone be in agreement? However, Young Earth advocates make these claims without recognizing that such assertions undermine their own arguments and represent a fundamental lack of understanding of how science works.

First, consider the general notion that disagreement is indicative of weakness. The corollary to this must be that *agreement* is a sign of strength. By this reasoning, Young Earth advocates would have to concede that ages in the hundreds of millions of years for the Grand Canyon layers must be valid, because agreement among scientists on this subject is virtually unanimous. Second, if disagreement over details is truly a reason to discount an entire scientific framework like uniformitarianism, then the flood geology framework itself fails in a deluge of self-contradicting arguments. Young Earth journals are full of disagreement on a host of subjects, including when the Grand Canyon was cut. On the subject of the Grand Canyon and Grand Staircase layers, Young Earth proponents do not even agree on which layers should be considered to be flood deposits and which should be considered to be post-flood deposits.

The practice of real-life science is much like figuring out the correct passageway through a maze. Not those printed mazes where you can see from the start where the exit lies, but one that you find yourself *inside of*, where the exit can only be found by experimenting with different routes. If you travel with a group, there may initially be considerable disagreement over which route is most likely to reach the goal. Competing hypotheses will be formed and members will test the different options, traveling some distance down different paths and reporting back on dead ends, promising new leads, or surprising developments (like strangely narrowing passages?). During the reporting process, others in the group have the opportunity to review the findings and identify errors (such as a missed passageway). It may even turn out that multiple routes lead to an exit, but also that one is much shorter or more easily traversed than the others.

Science works the same way. When data are initially coming in, there are many possible explanations to consider, and different researchers will disagree – sometimes passionately – about which is best. They put together arguments, test their hypotheses, and report on their work so that others can review, critique, support, or challenge them with new data. And in the process, everyone's understanding increases. Over time, general agreement is typically reached on broad questions, which leads to more detailed questions that require more data, more analysis, more dialog (agreement and disagreement), more review – and still greater understanding. Disagreement does mean there is some uncertainty in a particular detail under discussion, but working through disagreements is what ultimately leads to greater understanding and to increases in certainty.

For the Grand Canyon, there is broad agreement and a high degree of certainty on the timing and history of each rock layer's formation, with the oldest rocks dating back roughly halfway through the Earth's 4.5-billion-year history. Our understanding of exactly how these layers were carved to form the Grand Canyon is a work in progress, though much has been learned in the last 50 years – and all of it is increasingly at odds with the flood geology model.

Figure 17-6. Scientists investigating a cave in the Grand Canyon. *Photo by Debbie Buecher.*

The opening to Stanton's Cave in the Redwall Limestone, River Mile 32. *Photo by Tim Helble.*

CHAPTER 18

LIFE IN THE CANYON
PACKRATS, POLLEN, AND GIANT SLOTHS

by **Joel Duff**

Chapters 16 and 17 focused on the geologic processes responsible for carving the Grand Canyon. In this chapter, we will look at the plants and animals that began to occupy the canyon as it formed, with particular interest in what we find preserved in cave deposits. These deposits contain evidence of what kinds of plants and animals lived here in the past and provide additional clues about the canyon's age.

Remains of animal or plant life in a cave reflect the time after the cave became accessible from the surface. When a canyon begins to form in cave-bearing layers, erosion steadily opens up new caves and then steadily erodes them away. In the Grand Canyon, continual erosion along the main channel and its tributaries means that the earliest exposed and inhabited caves were eroded away long ago. Caves found today represent the most recent ones to be opened and inhabited, so it is not surprising that we

Figure 18-1. A partially fossilized packrat midden in a large cave crevice. *Photo by Debbie Buecher.*

Figure 18-2. Modern packrat midden. *Photo by Tim Helble.*

don't find millions of years of history preserved in them. But we do find tens of thousands of years of history — and evidence of very different life forms and environmental conditions compared with those that are present today.

Packrat Middens, Animal Droppings, and Pollen

Packrats get their name for a good reason — they are hardwired to live in nooks that contain not only their own collections of plant material, but also the leftovers of their ancestors' collections. The result is a layered nest, or midden, containing multiple generations of collected plant material (Figs 18-1, 18-2, 18-3). Individual packrats tend to limit their foraging to a distance of about 300 feet from their nests, so whatever is found in their nest likely grew very close by. Middens tend to preserve material well because of the combination of dry climate and periodic additions of rat urine (Fig 18-4). A number of packrat middens in Grand Canyon caves have been carbon-14 dated to be about 20,000 to 10,000 years old (older than the last glacial maximum period).

In larger caves that have been occupied by multiple generations of plant-eating mammals, deposits of semifossilized and unfossilized dung have been discovered — some of which are many feet thick. In packrat middens and in dung, abundant pollen

Figure 18-4. Author Wayne Ranney holding a piece of midden material cemented by packrat urine. *Photo by Tim Helble.*

is found that allows us to identify the specific plant types the animals were collecting and eating. This, in turn, tells us about the plants that were once growing in the vicinity. This record of past plant communities reveals a history of changing climate and a time when the Grand Canyon was much colder and wetter than it is today. For example, packrat middens and dung in a number of caves contain organic material that reveals long periods of time when the lower reaches of the canyon were populated by juniper and ash forests.

Despite the historical evidence of a well-developed cool-weather forest ecosystem, it appears that it was never witnessed by human eyes. All remains of plants and animals found in conjunction with human occupation of the canyon's caves occur in younger deposits, and all reflect the dry, warm climate that we are familiar with in the Grand Canyon today.

Extinct Giant Ground Sloth

Exploring one of the Grand Canyon caves, we find that we have been provided a unique window into a time in which many strange animals once roamed the southwestern United States. This cave is Rampart Cave, located in the western Grand Canyon more than 4,000 feet below the canyon rim. Rampart Cave is about 150 feet long, and when

Figure 18-3. The white-throated wood rat (*Neotoma albigula*) commonly known as a "packrat." These animals live in the Grand Canyon area today but are rarely found in caves where fossil packrat middens occur, because the canyon's vegetation has undergone major shifts as a result of thousands of years of climate change. *Photo by Ken Cole, U.S. Geological Survey.*

Figure 18-5. Skeletons of extinct giant ground sloths in the Smithsonian Museum of Natural History in Washington D.C. To the right is stuffed remains of an extinct giant ground sloth, "Gertie." The animal was over 15 feet tall and weighed about 2,000 pounds. This particular sloth lived about 11,000 years ago, as determined by carbon-14 dating. Note the long, sharp claws, suitable for digging. The sloths were vegetarians, as is evident from their dung. Gertie was found preserved in the entrance area of Grand Canyon Caverns and was stuffed for display in the cave. *Photos (left) Tim Helble, (right) Bob Buecher.*

Figure 18-6. Bin filled with sloth dung in Rampart Cave. The dung was put into bins during archeological investigations of the early 1970s. *Photo by Doug Powell.*

Figure 18-7. Close-up of sloth coprolite (poop) showing the type of vegetation eaten by the giant sloths. *Photo by Debbie Buecher.*

it was discovered its floor was covered with the dung of an extinct giant ground sloth, *Nothrotherium shastense* (Fig 18-5). Remains of this bear-sized sloth have been found across much of North America. The earliest human arrivals to the continent might have actually interacted with these magnificent creatures, but the sloths had been extinct for several thousand years before the first evidence of human activity in the Grand Canyon.

Radiocarbon (C-14) dating of fecal material on the floor of Rampart Cave has yielded a minimum age for the upper layers of more than 11,000 years, suggesting the cave has been uninhabited by sloths since that time. The youngest dung samples also contain the pollen of juniper and ash trees, which reflects the much cooler climate that ex-

isted when these giant sloths lived. Samples from the deepest dung deposits in the cave have yielded ages greater than 35,000 years, suggesting that the cave was occupied on and off by giant sloths for almost 25,000 years. There is no evidence that humans ever overlapped with the giant sloths in Rampart Cave – the earliest human artifacts in Grand Canyon caves (e.g., split-twig figurines shown below) have been radiocarbon-dated to about 6,000 years ago (Fig 18-8).

Problems for Flood Geology

The fact that giant sloths went extinct long before the arrival of humans in the Grand Canyon poses a major problem for the supposed dispersion

Figure 18-8. Split-twig figurine in a Grand Canyon cave. Figurine was photographed in the position in which it was found, and it was left untouched, as is the policy of Grand Canyon National Park. *Photo by Bob Buecher.*

of animals and humans over planet Earth from the ark's resting place in the mountains of Ararat in the distant Middle East. That is, the time of that dispersion (since all animals and people supposedly died in the flood except for Noah, his family, and the animals on the ark) would have had to occur in the last 5,000 years since the flood. Therefore, a flood geology model allows for virtually no time between the arrival in the Grand Canyon of large animals like giant sloths and the arrival of humans. So how does the flood model account for the large gap between the radiocarbon-dated ages of sloth and human remains? The only explanation is to argue for hyper-fast radiometric decay rates, yet we saw in Chapter 9 how this claim has been put to the test and found to be impossible. Caves with 35,000-year-old sloth dung is a strong testament against the flood model.

Viewed from a modern geology perspective, the piles of dung that contain plant material from a different climate and that significantly predate the first arrival of humans present no real challenges. The span of time represented by radiocarbon ages is consistent with extended habitation by generations of sloths, the plant material in their dung fits with a more temperate climate during the last ice age, and the subsequent extinction of these creatures is consistent with climate changes that made conditions inhospitable to their way of life.

Figure 18-9. A modern-day inhabitant of the Grand Canyon: a bighorn sheep. *Photo by Tim Helble.*

View of the eastern Grand Canyon. *Photo by Mike Koopsen.*

PART 5

A Verdict on Flood Geology

Part 2 focused on the tools geologists employ to sort out the Earth's history as recorded in rocks. Many examples were provided, but with no attempt to see how well those individual pieces fit together to tell the "grand story" of the Grand Canyon. This next step is critical, because explanations for this layer or that feature that seem to work well when considered independently may fall apart when considered as a whole.

As an example from daily life, it is quite plausible that someone saw you having lunch at a café in New Orleans and someone else saw you having lunch at a bistro in Portland. Both could easily be true. So let's say we try to combine these sightings into a single larger story of your life. If one sighting was on January 1 and the other on July 1, the two stories could fit the profile of someone who is well traveled and has a fondness for croissants. But now suppose that both sightings are claimed to have been at noon on January 1. Now the pieces do not fit, and something is wrong – you could not have been in two places at the same time. Since we are not willing to consider teleportation, the most probable explanation is that one person simply mistook a stranger for you in one of those cities. The same assessment can be applied to the various explanations offered for each rock layer in the Grand Canyon, as we consider how well the suggested histories fit together when viewed in sequence.

We started putting some of the pieces together in Part 3 by recognizing a pattern of life forms that appear in one set of rock layers and disappear in the next. We found patterns that fit well with one model (life forms appearing and disappearing over long stretches of time) but that do not fit with the other model (all varieties of life present at the same time and dying in a single catastrophe). Part 4 also considered how well multiple observations fit with flood geology or conventional geology models, though that investigation was limited to the more recent time of canyon carving.

Now we are finally ready to consider how explanations for each piece of the canyon's full history fit together into a single comprehensive story, as told by flood geologists and as told by conventional geologists. We will accomplish this with a 7-mile hike that starts with the oldest rocks down at the river and finishes with the youngest up at the rim. Along the way, we will point out characteristic features – such as rock types, layering, sedimentary structures, faults, and fossils – and we will draw from what we learned in previous chapters to try to understand how these rocks and features formed and fit together. The conclusions of conventional geologists will be compared with those of flood geologists to see how well each group's conclusions account for all the data, and, more importantly, how well they fit together into a composite story.

Hikers on the South Kaibab Trail, just above Kaibab Bridge. *Photo by Mike Koopsen.*

CHAPTER 19

RIVER TO RIM

PUTTING ALL THE PIECES TOGETHER

by **Gregg Davidson** and **Wayne Ranney**

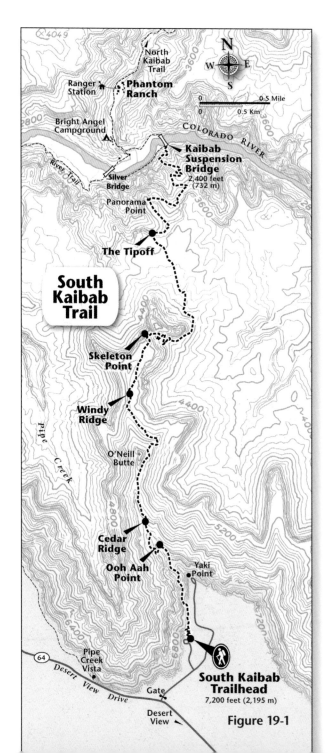

Figure 19-1

THE GRAND CANYON IS THE SINGLE BEST place in the world to observe the broad span of Earth history, and the South Kaibab Trail is one of the best places within the canyon to see its spectacular geology (Fig 19-1). The South Kaibab is a very popular trail that is easy to access by a free shuttle service from Grand Canyon Village. Most of this chapter will describe what can be seen specifically along this trail, but occasional references will be made to what is observed at the same level on other trails. The South Kaibab Trail is unusual in the Grand Canyon because it was created by

Figure 19-2. Wayne Ranney on the South Kaibab Trail. Ranney has taken more than 400 trips into the Grand Canyon, with at least 250 of those on the South Kaibab Trail. Many of the photos in this chapter are from Wayne's personal collection, spanning nearly four decades of exploration, photography, and writing about the canyon. *Photo by Helen Ranney.*

dynamiting into a high ridge rather than by following along the bottom of a side canyon, like most other trails in the canyon. For this reason, the views east and west from the trail are tremendous, and the canyon's geologic story is on dramatic display. But enough introduction – let's hike!

1. Vishnu Schist and Zoroaster Granite *(mile 0 to 0.4)*

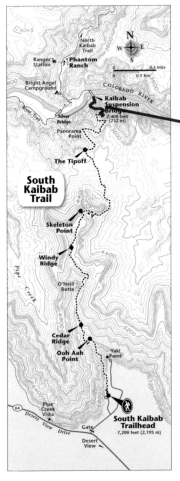

Our trip begins along the banks of the Colorado River near Phantom Ranch – a small resort and common hiking destination at the bottom of the Grand Canyon. The South Kaibab Trail starts at the Colorado River just south of the ranch, where the Kaibab Suspension Bridge crosses the river (Fig 19-3). Standing at the bridge (mile 0 in our hike), we are surrounded by the Vishnu Schist, a rock riddled with bands of pink Zoroaster Granite (Fig 19-4, and page 186 photo). A few quick observations tell us much about the history of this rock. The schist contains altered minerals that form under high temperatures and pressures typically found only at great depths – on the order of 10 miles or more beneath the surface. These rocks are also folded and contorted. Rapid bending and folding shatters rock, but the schist shows little evidence of shattering – which indicates that it deformed very gradually. The granite that crisscrosses through the schist contains large crystals, which indicates that the rate of cooling was also slow (Fig 19-5).

All of these features are consistent with rock that formed over a long period of time, deep below the surface – much deeper than where this rock is exposed today. Samples of the crisscrossing granite have been radiometrically dated to about 1.7 billion years old. The granite intrudes into the schist, which means the age of the schist – the time when sedimentary and volcanic rocks were buried deeply enough to experience metamorphism and alteration to schist – is even older.

Flood geologists claim a recent and rapid origin for these rocks, with formation, intrusion, alteration, and cooling all between the creation week and Noah's flood (only about 1,650 years). But none of this fits with what we see in the rocks. From lab experiments, we know it takes a tremendous amount of time for large masses of heated and deeply buried rocks to cool. Given the immense volume of rock in the Zoroaster Granite, a recent pre-flood origin

Figure 19-3. Kaibab Suspension Bridge. *Photo by Tim Helble.*

Figure 19-4. Wavy bands in the Vishnu Schist. *Photo by Wayne Ranney.*

Figure 19-5. Large crystals in the Zoroaster Granite. *Photo by Debbie Buecher.*

Figure 19-6. South Kaibab Trail descends through Supergroup layers (black arrow, Hakatai Shale; yellow arrow, Shinumo Quartzite), but upstream, the Supergroup layers are absent (white arrow). *Photo by Gregg Davidson.*

would have required the action of some unknown mechanism to achieve ultra-rapid cooling. Flood geologists also claim that radiometric dates are unreliable. But no alternative for dating these rocks is offered – other than one based, not on the Bible, but on a particular interpretation of the Bible.

Remember that flood geologists say natural processes can account for all the Earth's layers that formed after the close of the creation week, yet the very start of the Grand Canyon's story requires reliance on never-before-encountered natural mechanisms and ignores all evidence of great age. We'll revisit this subject again later in our hike.

2. Grand Canyon Supergroup and the Great Unconformity *(mile 0.4 to 1.9)*

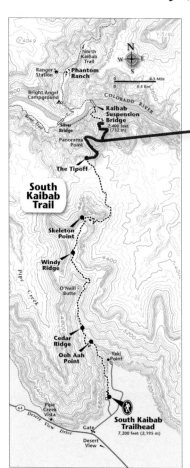

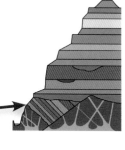

If you are exploring the geology of the Grand Canyon for the first time, the rocks along this stretch of the South Kaibab Trail might seem confusing. For the next mile, we pass through tilted rock layers belonging to the Grand Canyon Supergroup. Along the trail we see three formations: the Bass Formation, Hakatai Shale, and Shinumo Quartzite, with the Vishnu Schist below and the horizontal Tapeats Sandstone above (Fig 19-6). However, if we look up or down the river canyon, we see cliff faces where the Vishnu Schist terminates directly against the Tapeats Sandstone – the

Figure 19-8. Stromatolite layers. *Photo by Wayne Ranney.*

Figure 19-7. *top:* Hotauta Conglomerate at the base of the Supergroup, containing pieces of schist and granite. *bottom:* Cut and polished Hotauta Conglomerate. *Photos by Wayne Ranney.*

Supergroup layers are entirely absent. There is clearly a complex history here that would require a lengthy discussion to fully describe. For the purposes of this chapter, we will limit our observations to those most pertinent to comparisons between the expectations of conventional geology and flood geology.

The Grand Canyon Supergroup includes nine formally named rock formations, with a combined total thickness in excess of 12,000 feet, but we find only the lowest three formations along the South Kaibab Trail. In some parts of the canyon (such as along the New Hance Trail), the lowest Supergroup layer contains chunks of the underlying schist and granite (Fig 19-7). The presence of such chunks reveals much about the history at this location. A very specific sequence of events was required in

order for pieces of schist and granite to be encased within the lowest Supergroup layer: the crust had to be uplifted, miles of overlying rock had to be eroded away to expose the schist at the surface, fragments of the eroded rock had to be collected in low spots, and, finally, sediments had to be deposited over and around the weathered chunks of schist and granite. None of this requires any unusual processes to accomplish – unless you need it all to have happened in just a few hundred years, without the benefit of a great cataclysm (recall that flood geologists identify all these rocks as pre-flood).

About a mile up the trail in the Bass Formation, we encounter our first visible fossils – colonies of single-celled organisms called stromatolites (Fig 19-8). No multicellular fossils are found in the Bass Formation or in any of the overlying Supergroup layers – nor do we find multicellular fossils in rock layers of similar age anywhere in the world. If multicellular life did not exist at the time these deposits formed, what we find makes perfect sense. On the other hand, if the pre-flood Earth was filled with all manner of life forms that were similar to those we see today, and the Grand Canyon Supergroup was deposited before the start of the flood, at least some of these layers should contain a mix of all these types of life.

We mentioned earlier that the view up or down the river corridor here yields a very different view from the one along the trail. The Supergroup layers up and downstream are entirely missing, and the Tapeats Sandstone is sitting directly on the Vishnu

Schist (Fig 19-9). The dramatic difference is a testament to the work of faulting. This section of our trail crosses two faults, the Cremation and Tipoff Faults. During past tectonic activity, the block of rock between these faults, including layers of the Supergroup, slid downward several hundred feet and came to rest below the surrounding rock. Later erosion removed all of the Supergroup layers from the higher adjacent blocks, exposing the underlying schist and granite. The block crossed by the South Kaibab Trail (between the Cremation and Tipoff Faults) sat much lower, shielding the lowest Supergroup layers from erosion. Later, renewed deposition covered the region with blankets of new sediments, starting with the Tapeats. A simplified sequence of events is illustrated in Fig 19-10.

Once again, normal Earth processes can readily explain these observations, with shifting tectonic plates rupturing crust and moving blocks upward and downward. No never-before-seen mechanisms or mysterious forces are necessary to create the rock formations that we see today. The same cannot be said for the flood geology viewpoint. Most flood geologists place the deposition of all these rocks between Day 3 of the creation week and the flood. The tilting and faulting is attributed to violent tectonic activity initiated at the start of the flood, erosion of entire blocks of the Supergroup to mega-tsunamis, and deposition of the overlying layers to ensuing flood surges. Inconsistencies abound. We'll consider four topics that highlight these inconsistencies.

Thickness

Flood geologists claim that Noah's flood deposited all of the flat-lying Grand Canyon strata that sit above the Great Unconformity, including the Tapeats Sandstone and much of the Grand Staircase

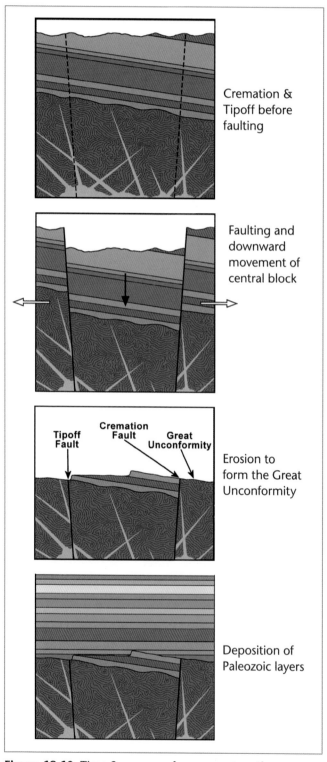

Cremation & Tipoff before faulting

Faulting and downward movement of central block

Tipoff Fault Cremation Fault Great Unconformity

Erosion to form the Great Unconformity

Deposition of Paleozoic layers

Figure 19-10. Time Sequence of movement on the Cremation and Tipoff Faults, erosion, and deposition.

Figure 19-9. Vishnu Schist–Tapeats Sandstone contact, Supergroup missing. *Photo by Wayne Ranney.*

rocks. These strata have a combined thickness of about 15,000 feet. If it required a cataclysmic flood to deposit 15,000 feet of sediment above the Great Unconformity, how did over 12,000 feet of Supergroup sediments accumulate before the flood? Is a second, earlier catastrophe required? Similarities between the layers above and below the Great Unconformity mean that either two global cataclysms occurred at different times (the first of which is missing from the biblical record), or that layers both above and below formed by normal Earth processes.

Erosion along top and bottom of Supergroup

Flood geologists argue that a violent global flood eroded away rock at the top of the Supergroup to form an unconformity. Yet the erosional surface between the Vishnu Schist and the base of the Supergroup is not fundamentally different from the one at the top of the Supergroup. Why is a global catastrophe needed to explain the unconformity at the top of the Supergroup but not the equally extensive unconformity at its base? Also, how was the schist beneath the Supergroup rocks uplifted and eroded before the flood? Recall that erosion first requires that the rock be lifted up so that it can be worn down. According to flood geologists, plate tectonics began when the fountains of the deep burst open and the Earth's crust was set violently into motion. With no mechanism to uplift the schist, the only option is to stuff it all into Day 3 of the creation week, when land was separated from the waters. Translation: the only plausible explanation is that everything in the schist, plus its uplift and erosion, is of miraculous origin.

Sediment types and structures

The alternating layers of conglomerate, limestone, shale, and sandstone in the Supergroup — including features like ripple marks, mud cracks, and cross bedding — look remarkably similar in nature to the alternating layers of conglomerate, limestone, shale, and sandstone in the overlying Paleozoic layers. The only substantial difference is the type of fossil organisms found. Why would layers deposited by a global flood look so similar to layers deposited by normal processes? Sediments and specific sedimentary features that are said to support a violent flood history for the Paleozoic layers appear convincing only by ignoring the presence of the same types of sediment and features in the Supergroup layers.

Fossils (or the lack thereof)

According to flood geologists, death began at a specific point in time between the creation week and Noah's flood, only a few thousand years ago. The first layers that contain visible fossils are in the Supergroup Bass Formation (of the Unkar Group), so all layers there and above represent deposition while organisms were dying. According to Genesis, Chapter 1, all the major categories of modern organisms were present prior to the flood, which means layers in and above the Bass Formation should contain a representative sampling of life forms reflective of the modern array

Figure 19-11. Tapeats Sandstone, containing boulders of Shinumo Quartzite. Arrows point to quartzite boulders, roughly 10-12 inches in size. *Photo by Wayne Ranney.*

of organisms. Many of these rock layers appear to have been deposited in marine settings, so at the very least, we should find fossils of shelled organisms, fish, coral, lobsters, and an occasional marine reptile or mammal. But in the entire 12,000 feet of Supergroup rock, there are no complex organisms at all. Not a single fish, clam, snail, coral, tooth, or bone. Fossilization was obviously occurring, so how did all but single-celled varieties escape preservation – over the entire planet?

3. An Eroded Cliff along the Great Unconformity *(mile 1.9 to 2.0)*

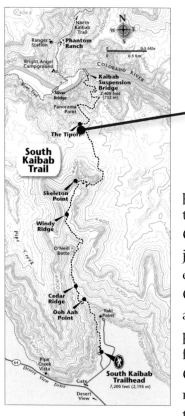

At the end of the previous section of the trail, we reached the Great Unconformity – just before the trail emerges from the Inner Gorge and begins to approach the wide expanse of the Tonto Platform. At this spot, the Great Unconformity is represented by the contact between the Supergroup's Shinumo Quartzite and the overlying Tapeats Sandstone. But the contact here is not horizontal. In fact, the Great Unconformity at this point becomes nearly vertical, with Shinumo Quartzite on one side and the Tapeats Sandstone on the other. Upon closer examination, we find that large angular blocks of the quartzite are encased within the sandstone, forming a conglomerate at the base of the Tapeats (Fig 19-11).

These observations testify to a time when the Tapeats Sea was encroaching on land and eroding into a cliff face of Shinumo Quartzite. Blocks of quartzite periodically fell into the water and lodged in soft

layers of sand, accumulating at the base and eventually being covered by new deposits of Tapeats sand. The angular shape of the quartzite blocks means the quartzite was hard at the time of erosion of the cliff face, and we can see evidence of disrupted sand directly beneath some of the fallen blocks. Eventually, sea level rose high enough to deposit sand over the entire cliff and accumulate rubble of Shinumo Quartzite boulders.

At first pass, this history of crashing waves and rapidly buried chunks of broken rock could be argued to fit within the flood model, but only if viewed in isolation from all the underlying and overlying layers that have their own histories – histories that do not fit well with a single catastrophic event.

4. Tonto Group: Tapeats Sandstone, Bright Angel Shale, Muav Limestone *(mile 2.0 to 3.0)*

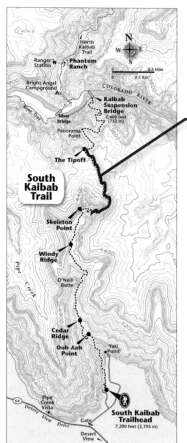

At the top of the Tapeats Sandstone, the slope suddenly becomes more gradual. If hiking down from the top, a long stretch of gently sloping trail abruptly plunges downward at this spot – aptly named *The Tipoff*. Coming up the trail, we pass The Tipoff and step onto the Tonto Platform – a broad ledge hosting panoramic views of the canyon (Fig 19-12).

Figure 19-12. Tonto Platform, viewed from the South Kaibab Trail. *Photo by Wayne Ranney.*

Fig 16-4, page 166 for a reminder of how this process works).

Farther up, the South Kaibab Trail again inclines steeply through the debris-covered slopes of the remaining Bright Angel Shale and the cliff-forming Muav Limestone. The Tapeats Sandstone, Bright Angel Shale, and Muav Limestone, all of Cambrian age, together comprise the Tonto Group. These layers share a natural association, because the transition from sand to silt and clay and then to limestone fits well with what we would expect from a gradually subsiding coastline or rising sea (see Fig 5-13, page 62).

Looking a bit closer, we see that the transition from shallow to deep water was not uniform. Though the overall change from sandstone to shale to limestone is obvious, there are numerous alternating layers that identify multiple episodes of small-scale increases and decreases in relative water depth (Fig 19-13). If we were to venture off the trail and follow the layers westward, we would find grain sizes becoming finer, consistent

Figure 19-13. Alternating layers of the Muav Limestone (seen from the Tanner Trail). *Photo by Gregg Davidson.*

The platform owes its existence to the exposure of the easily eroded Bright Angel Shale. The shale's relatively fast rate of erosion has undercut the harder layers of Muav Limestone above it, causing collapse. It is these alternating layers of hard and soft rock that give the Grand Canyon its classic alternating cliff-and-bench profile (see

Figure 19-14. Fossil worm trails in Bright Angel Shale. *Photo by Gregg Davidson.*

Figure 19-15. Trilobite in Bright Angel Shale. *Photo by Mike Quinn, Grand Canyon National Park.*

with deepening water in that direction. From this information, we can tell that the shoreline was advancing toward the east as sea level rose, with many back-and-forth fluctuations over shorter periods of time.

Fossils found in these layers are typical of those of the Cambrian Period in many other parts of the world. Forty-seven different species of trilobites have been identified in the Tonto Group alone, but none of these species are found in any layer above or below the Tonto Group (Figs 19-14, 19-15). Multiple layers

of tracks and burrows provide evidence of a long succession of established ecosystems, one on top of the other. The fact that organisms such as trilobites appear and disappear from the Grand Canyon fossil record in the same order as they do in strata around the rest of the world tells us that each layer of the Grand Canyon represents a distinct time period in Earth history.

As with the flood geologists' explanations for the Supergroup, there are many problems and inconsistencies with their explanations for the Tonto Group. We'll once again consider just a few examples.

Sorting

Particles can become arranged by size (sorted) when flowing water gradually slows down; larger particles begin settling out first and smaller ones later. Flood geologists see the general sorting of particles to form sandstone, shale, and limestone as evidence of the floodwaters deepening and slowing. Their argument assumes that the limestone at the top of the Tonto Group did not originate from the shells of organisms (as with most limestone formation), but was washed in as lime sediment from some distant source. Even if we ignore the improbability of lime particles surviving transport without dissolving (see Chapter 5), there is still a serious problem. The flood model also assumes that the lime particles must have been exclusively smaller (or less dense) than the clay particles for them to settle out into separate, relatively unmixed layers as the water slowed. But, in actuality, the size and density of lime and clay particles overlaps substantially, meaning that the flood model should expect these layers to be *highly* mixed. When we look within the layers of shale and limestone of the Tonto Group, the degree of mixing is very small, contrary to expectations from deposition by a great flood.

In other words, none of the flood-geology arguments for layering are supported by the actual evidence found in the Grand Canyon.

Limestone

No flood of any size, including the colossal floods of the Channeled Scablands, has ever been found to leave behind limestone or salt deposits around the world. The explanation that these deposits arose from the release of hot, mineral-rich fluids from the floodgates of the deep requires that separate pockets of calcite-rich and salt-rich subterranean waters were created during Day 3 of the creation week (when the land was separated from the waters) and were set in reserve for the coming flood. Upon release, separate calcite-rich and salt-rich fluids had to somehow avoid mixing with each other, avoid dilution by mixing with seawater, and form mineral precipitates without mixing with sediment stirred up by the churning flood waters – each step requiring a separate miracle. Limestone precipitation likewise would have required a miracle, for any hot, mineral-rich water released into the ocean would have *dissolved* calcite as it cooled – limestone could not have precipitated out of the water without violating basic laws of chemical thermodynamics.

Fossils

The ordering of fossils could be discussed with any number of fossil types, but here we'll limit the conversation to trilobites. The 47 species of trilobites known from the Tonto Group come in a variety of sizes and shapes. As ocean-dwelling creatures, trilobites should have had widespread distribution and are, in fact, found in Cambrian layers all over the world. How did every variety, large and small, stubby and elongated, get sorted into the same group of layers in the same sequence around the world, without a single case of mixing with a jawed fish? And why don't at least a few of these 47 species occur in layers between the Cambrian and the canyon rim, if all of these layers were deposited in the short duration of 150 days?

5. Temple Butte Formation (mile 3.0)

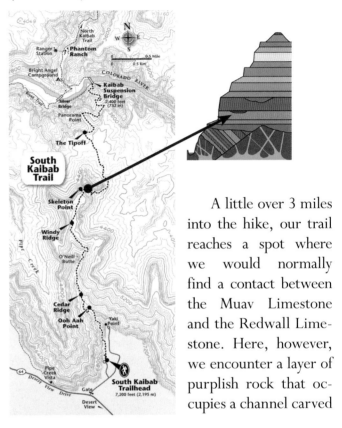

A little over 3 miles into the hike, our trail reaches a spot where we would normally find a contact between the Muav Limestone and the Redwall Limestone. Here, however, we encounter a layer of purplish rock that occupies a channel carved

Figure 19-16. A pocket of Temple Butte rock along the South Kaibab Trail. *Photo by Wayne Ranney.*

out of the top of the Muav (Fig 19-16). This is an outcrop of the Temple Butte Formation. This formation is not found everywhere in the canyon, and it varies considerably in thickness. It gets its own name because it contains a unique assemblage of fossils compared with those of the Muav or Redwall, and it sits in low spots (channels) carved into the top surface of the Muav. Recall that Muav fossils are typical for the Cambrian Period. If channels were carved out soon after deposition and then quickly refilled, we should expect more Cambrian fossils, or at least fossils from the following Ordovician period. The fossils actually found, however, such as placoderm fish and various corals, come from the much later Devonian Period. This tells us that some time after the Muav Limestone formed, erosion removed whatever layers might have been deposited above, eventually scouring channels down into the Muav surface. Still later, new deposits filled in the channels. The total gap between deposition of the Muav and deposition of the Temple Butte represents about 135 million years.

Flood geologists call upon submarine currents during the flood carving channels into soft Muav lime, which subsequently refilled with fresh lime sediment, all within a few days. Under such a scenario, why would a unique set of organisms (all Devonian), including both bottom-dwelling corals and free-swimming placoderm fish, settle out exclusively into these channels and nowhere else?

6. Redwall Limestone
(mile 3.0 to 4.0)

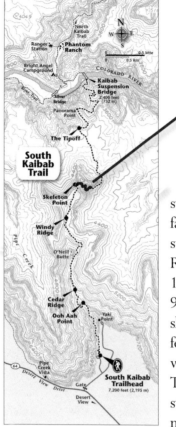

The trail gets quite steep as we ascend the famous "Red and White" switchbacks through the Redwall Limestone (Fig 19-17). The Redwall is 98% pure calcite, with sheer cliffs reaching 500 feet in thickness. Together with the Muav and Temple Butte Limestones, these layers form massive cliffs that can rise more than 1,500 feet. Although the Muav, Temple Butte, and Redwall are all limestones, the contacts between them are striking because of the dramatic changes in fossil organisms. In the Redwall we find fossils of bony fish, shark teeth, and other organisms typical of the Mississippian Period. A closer look at the Redwall reveals that much of it is composed of crinoids — marine animals that resemble flowers with tentacles — that are attached to the ocean floor by long segmented stalks or stems. Because some crinoid species still exist, we know that when they die, the segmented stems easily come apart and litter the ocean floor with their remains (Fig 19-18).

The Redwall Limestone extends laterally into several states beyond Arizona (given different formation names

Figure 19-17. Switchbacks through the Redwall Limestone. *Photo by Wayne Ranney.*

Figure 19-18. Crinoid fragments in the Redwall Limestone, and live crinoid from the Solomon Islands. *Upper left photo by Wayne Ranney and photo insert (right) by Tim Helble.*

in other places). Such an expansive layer of crinoid remains, with virtually no intermixed sand or clay, indicates a vast, shallow, warm sea with submarine colonies of crinoids blanketing the seafloor. The absence of any significant clay or silt means that the crinoids must have lived far from where sediment-laden streams discharged to the ocean. This thickness of fossil remains is easily accounted for by a succession of generations of crinoids, each growing on top of the remains of older ones. Similar processes are at work today where modern reef systems grow on top of the remains of their ancestors and produce limestone hundreds of feet in thickness.

Flood geologists argue that the crinoid stems in nearly pure calcite formed as the result of hot, calcite-rich fluids being released from the fountains of the deep and arriving at the same time that the crinoids were stripped from a distant ocean floor and

dashed into pieces, ultimately to be mixed with the precipitating calcite.

Imagine what is required by this explanation. At this stage in Noah's flood, mega-tsunamis have been ravaging the planet, scouring sediment and transporting it vast distances to accumulate in beds hundreds of feet thick across what is now the American West. Mineral-rich water, released when the tectonic plates broke apart (the "floodgates of the deep"), had to travel across the continent to this location without mixing with seawater (cooling and mixing would work against precipitation) and without mixing with any churned-up sediment. Meanwhile, gargantuan submarine communities of fragile crinoids, covering thousands of square miles, had to survive the early weeks (or even months) of violence without being ripped up or buried, only to be abruptly torn from the seafloor and transported hundreds of miles – also without mixing with any churned-up clay or sand. Finally, the undiluted mineral-rich water had to meet the flow of crinoid stems (free of mixed silt or clay) and precipitate around the crinoids as virtually pure calcite.

As unbelievable as this scenario is, there is more. If a single generation of crinoids living at the time had been buried in place, the disseminated parts would have coated the ocean floor with a few inches of crinoid remains. To get deposits 500 feet thick over multiple western states, the crinoid communities would have had to initially cover a vastly larger area – tens of thousands of square miles in size – that were ripped up and stacked within a smaller area – all in clear, un-muddied water.

7. Surprise Canyon Formation
(mile 4.0; not exposed on the South Kaibab Trail)

Figure 19-19. Surprise Canyon Formation, viewed from the Colorado River at the mouth of Fern Glen Canyon. *Photo by Wayne Ranney.*

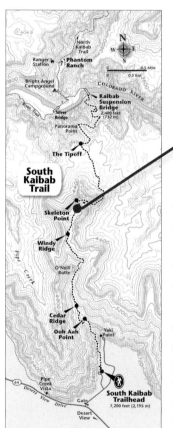

At the top of the arduous ascent up the trail in the Redwall Limestone, four miles from the river, we come to the contact with the Supai Group. In other parts of the canyon, however, another formation is nestled in low spots on top of the Redwall – the Surprise Canyon Formation (Figs 19-19, 19-20). When the low spots are mapped, they form connecting conduits that look like a stream network, with channels that become wider and deeper in a westerly direction. The largest reaches a depth of 400 feet.

The bottom layers of the Surprise Canyon Formation contain fragments of the underlying Redwall Formation, complete with fossils from the Redwall – a clear sign of exposure and erosion of the Redwall surface following retreat of the ocean. Sediments filling these channels contain a rich assortment of fossils. In the lower sediments, the fossils are of land-dwelling organisms, including trees. Such information speaks of uplift, or of falling sea level, that left the area above

Figure 19-20. Fragments of the Redwall Limestone in a conglomerate boulder from the Surprise Formation. The boulder is about 1.5 feet wide. *Photo by Erin Whitakker, National Park Service.*

water long enough for stream systems and terrestrial ecosystems to develop before the region was again submerged. The fossils here are distinct from what is found in the underlying Redwall and are typical of organisms from the Late Mississippian Period.

Many Young Earth authors and speakers simply overlook the Surprise Canyon Formation. Others acknowledge its presence, but fail to mention its abundant terrestrial fossils. One might possibly imagine some land organisms being swept into the ocean and settling in a low spot or two, but the prevalence of these fossils in the bottom deposits of the Surprise Canyon layers – without marine fossils mixed in – speaks loudly of a prolonged period during which rivers coursed through these channels and land ecosystems thrived.

Figure 19-21. Alternating layers of the Supai Group and the Hermit Formation, above the vertical Redwall Limestone and beneath the vertical, lighter-colored Coconino Sandstone. *Photo by Wayne Ranney.*

8. Supai Group and Hermit Formation *(mile 4.0 to 5.8)*

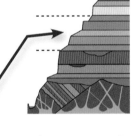

There is a brief respite from steep hiking past Skeleton Point, but soon the trail climbs rapidly again through the Supai Group and the Hermit Formation. Along the way we find alternating layers of sandstone, shale, limestone, and conglomerate (Fig 19-21). Myriad alternating layers are again consistent with many cycles of rising and falling sea levels, interspersed with periods of deposition and erosion. Filled-in low spots, like

Figure 19-22. Edge of channel in the top of the Supai Group, filled in with Hermit Formation sediments (at the 1.5 mile Rest House, Bright Angel Trail). *Photo by Wayne Ranney.*

Figure 19-23. Channel within the Supai Group (in the Wescogame Formation) near 27 Mile Rapid in Marble Canyon. *U.S. Geological Survey photo by E.D. McKee.*

Figure 19-24. Reptile trackway in the Supai Group. *Photo by Wayne Ranney.*

Figure 19-25. Fern fossils in the Hermit Formation. *Photo by Michael Quinn, National Park Service.*

those described for the Temple Butte and Surprise Canyon Formations, also appear where Hermit Formation sediments fill in ancient channels in the upper sandstones of the Supai Group (Fig 19-22). Additional channels are found within formations in the Supai Group (Fig 19-23). Some of the sandstone layers of the Supai also show cross bedding that is characteristic of wind deposition.

The first occurrences of tracks from vertebrate animals (those with back-bones) are found in the Supai Group (Fig 19-24). Some rock horizons in the Hermit Formation contain terrestrial fossils, such as dragonfly wings (Fig 13-12, page 139) and fern plants (Fig 19-25). The fact that terrestrial and marine fossils are

Figure 19-26. *right:* Coconino Sandstone-Hermit Formation contact (Bright Angel Trail). The Coconino is the lighter rock above the reddish Hermit rock. *Photo by Gregg Davidson.*

not found intermingled within the same layers is clear evidence of distinct intervals of time when the region was sometimes above sea level and other times below it.

Flood geologists insist that all these layers are marine in origin — not because all the layers actually contain evidence of deposition in a sea, but because their model requires it.

9. Coconino Sandstone
(mile 5.8 to 6.5)

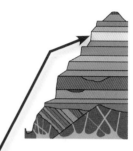

After the trail climbs through the red mudstone of the Hermit Formation and passes through the picturesque scrub trees along Cedar Ridge, it crosses into the Coconino Sandstone. A sharp unconformity is easily seen here, where the red beds below give way to the white Coconino above (Fig 19-26).

Figure 19-27. Reptile tracks in the Coconino Sandstone.
Photo by Wayne Ranney.

Not far above that contact is a boulder that contains the trackway of a small reptile (Fig 19-27). Here we can see the detailed preservation of each footprint – pad and claw marks – as well as avalanche impressions imparted as the animal's weight pushed on the loose sand. Trackways are much more important than single footprints, for they give information about the stride, weight, size, and gait of the animal. To the west, along the Hermit Trail, tracks are found that were left by scorpions and spider-like creatures. These tracks, coupled with the absence of any fossils of bone, speak of a time when desert sands covered the local landscape. Few organisms would have been present in such an environment, and conditions would have been unfavorable to the preservation of skeletons.

On ahead, the trail winds steeply up Windy Ridge – a spectacular section of trail that falls away on both sides. From here, we have a sense that we are nearing the top, though it will take a good measure of sweat to finish the trip. Cross bedding is abundant in the mostly quartz-sand layers. The etched grains and

the cross-bed angles reaching 30-34° are typical for a desert, sand-dune environment. The abrupt transition in color and composition between the Coconino Sandstone and the underlying Hermit Formation, together with fissures within the Hermit that are filled with sand, suggest a period of erosion down into the Hermit, with weathered cracks later filling in with sand from advancing dunes (more easily seen along the Bright Angel Trail; Fig 19-26).

For flood geologists, the flood was still raging at this time, so to them the Coconino *must* be a marine deposit. The tracks are said to have formed underwater, and tank studies with living amphibians are cited as evidence that tracks can be formed this way. The flood geologists fail to mention, however, that these studies were done in water with a weak current, yet a violent flow would be needed to transport the massive layers of sand that their model requires. Even modest flow velocities would quickly erase underwater tracks, such as the reptile and arthropod varieties found in the Coconino.

10. Toroweap and Kaibab Formations *(mile 6.5 to 7.0)*

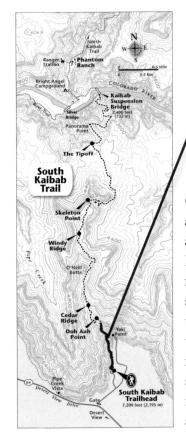

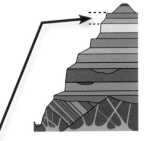

Near the top of the Coconino Sandstone, a spectacular view awaits us at Ooh-Aah Point – a worthwhile rest and admiration stop (Fig 19-28). Once we resume our hike, we climb through alternating limestone, shale, sandstone, and evaporites of the Toroweap and Kaibab Formations.

Figure 19-28. View from Ooh-Aah Point. *Photo by Wayne Ranney.*

DOES A DASH OF MARINE SEDIMENT MAKE THE WHOLE DEPOSIT MARINE?

Some isolated zones of *dolomite* (limestone with lots of magnesium – typically a marine deposit), have been found in the Coconino Sandstone. Flood geologists insist that the presence of dolomite means the *whole* system is marine. Given the proximity of the Coconino dunes to an ancient sea, it would not be surprising for some of the dune sands near the coastline to become cemented with dolomite. However, to say the entire Coconino is marine because of a pocket of dolomite is like finding a Swede living in Tokyo and declaring that all Japanese must be of Nordic stock!

The variable durability of the different layers produces a mixed terrain of steep and gradual slopes. The alternating layers reflect the return of cyclic sea level advance and retreat. With roughly half a mile to go, we encounter layers of rock that have tepee shapes, most likely created by the evaporites underneath buckling the rock upward (Fig 19-29). Such behavior is common when the weight of overlying rocks squeezes pliable underlying evaporites, like salt or gypsum, up through weak spots in the overlying strata. Evaporites provide strong testimony to periods of exposure during which intensive evaporation and drying occurred – no other mechanism is known to produce evaporite deposits. Fossil types change considerably as we move up from one set of layers to the next, which tells us that significant time must have passed while these layers formed – enough time to allow new ecosystems to replace older ones.

For flood geologists, the uppermost layers of the canyon bring us through only the first half of the flood (150 days, see Fig 3-2, pages 32-33). Though the flood is said to have been extraordinarily violent, sediment layers repeatedly formed with little mixing of different types or sizes of particles. Evaporites mysteriously formed underwater, and entire classes and orders of organisms waited until the second half of the flood to all be buried together.

Figure 19-29. Tepee structures in the Toroweap Formation. *Photo by Wayne Ranney.*

11. The Rim *(mile 7)*

Arriving at the rim, typically with aching legs and a great feeling of accomplishment, we have a fresh appreciation for the great depth of this canyon (Fig 19-30). As we catch our breath, we'll relate some final thoughts on what we have seen. The canyon's many layers, structures, and faults certainly represent powerful forces at work, but each is easily accounted for by normal Earth processes – some slow and some fast – but all normal. More importantly, the explanations for each individual layer or feature fit together into a larger story of rising and falling sea levels, and of slowly shifting tectonic plates lifting and lowering the crust. Fossils encountered along the trail in the canyon, and found around the world, communicate a consistent story as well – a story that makes sense only if the types of organisms present varied considerably at different times in the Earth's history. The fact that not a single fossil bird, dinosaur, mammal, or flowering plant can be found anywhere along this 7-mile hike is of great significance.

Flood geology arguments often have a ring of plausibility to them when they are applied to one layer or one feature in isolation, but there is no way to piece together all the individual explanations into a coherent whole. Flood geologists insist that God employed natural processes, open to scientific investigation, to lay down the Earth's myriad layers, but their explanations repeatedly require reliance on never-before-seen and mutually exclusive mechanisms. Radioactive decay had to be faster in the past, which would have required miraculous removal of heat to avoid melting the crust. Rapid plate tectonics likewise would have needed a miraculous

Figure 19-30. View from the South Rim at the South Kaibab Trailhead. *Photo by Wayne Ranney.*

dissipation of frictional heat. Mineral-rich fluids bursting through fissures at the start of the flood had to remain unmixed with seawater or mud or sand, and then violate basic laws of chemistry to precipitate limestone from a cooling fluid. Oriented fossils like nautiloids in one location are said to testify to deposition by a global flood, but nonoriented fossils everywhere else (which is the norm) are somehow not considered to be evidence of deposition in calm waters.

Earthquakes are called upon by some to explain separation of flood sediments into thin layers of differing particle sizes, while somehow allowing conglomerates to stay totally unsorted and fine features in delicate animal tracks to be unscathed. The immense record of fossil life is said to be evidence of a global flood that swept across entire continents, yet that flood somehow failed to capture a single mouse, seagull, whale,

frog, tulip, or lobster in the entire Grand Canyon sequence.

The conventional geologic understanding of the Grand Canyon is not just better than the flood geology view. The conventional model works; the flood model does not.

WHERE DID ALL THIS COME FROM?

All explanations by flood geologists are said to be based on the Bible. So where in Scripture do we find references to Noah's flood linked with earthquakes, shifting continents, rising mountains, tsunamis, and mineral-rich ocean vents? The number of verses is much like the number of bird or dinosaur fossils found in the canyon. The answer is zero. Exactly how, then, is flood geology a biblical model?

Vishnu Temple seen from Grandview Point at sunset. *Photo by Bronze Black.*

CHAPTER 20

SCIENCE VS. FLOOD GEOLOGY

NOT JUST A DIFFERENCE IN WORLDVIEW

by **the Authors**

AT THE BEGINNING OF this book, it was noted that Young Earth or flood geology adherents often claim that we are all looking at the same data but that our different worldviews cause us to "see" the data as evidence for vastly different and conflicting processes. The underlying assertion is that we are all practicing good science but are arriving at different interpretations of the data because of the biblical or humanistic "glasses" each person wears. Adherence to the Bible is said to draw attention to the natural evidence that supports an Earth of a very limited age, and rejection of the Bible causes one to see only the natural evidence of great age. However, the preceding chapters illustrate something quite different.

For each subject addressed in this book, when the data are considered in their totality and allowed to take us wherever they lead — without foreknowledge of the answer or a predetermined outcome — we are invariably led to a history of the canyon that extends back

Figure 20-1. Hiking toward Chuar Butte along the Beamer Trail. *Photo by Gregg Davidson.*

millions of years. A recent age for the canyon can only be imagined by deciding on such an answer in advance, carefully selecting bits of data that can be construed to fit the preconceived model, and ignoring data that do not fit. Herein lies the difference between science and flood geology — science goes where the data leads, flood geology does not. By deciding in advance what

Figure 20-2. Esplanade Platform in the western Grand Canyon. *Photo by Wayne Ranney.*

the answer to the question will be, flood geology does not study nature to discover what processes have been at work or what events may have transpired (or for that matter, what God actually did). Rather, flood geology starts with an answer and studies nature only to find those ways that fit with the predetermined model. In this respect, flood geology is the antithesis – the very opposite – of science.

The debate does involve separate worldviews, but not in the way flood geology advocates describe. True science is practiced by those whose worldview stipulates that nature is understandable, that the processes at work today on planet Earth can be used to inform us of what may have happened in the past, and that the fundamental laws of physics and chemistry have

not, and will not, change over time. These views are not inherently secular, for they developed under the historical influence of Christian philosophy, where science was deemed possible because of the guidance and sustenance of nature by a logical, consistent, and unchanging God. Science is thus practiced by the religious and nonreligious alike.

Contrary to the doctrine of flood geologists, the worldview of flood geology is not distinguished from other worldviews by its adherence to the scriptures found in the Bible. Rather, its distinguishing characteristic is adherence to a particular way of interpreting select passages within the Bible – accepted as fact, without considering any conflicting evidence within or outside the Bible. As a result, all data from nature must be force-fit into the accepted-truth model, no matter how convoluted the resulting story may become.

The message of flood geology is that what is observed in nature today cannot be used to inform us of what may have happened in the past, that fundamental laws of physics and chemistry cannot be assumed to be well understood, and – critically – that nature cannot be trusted to tell its own story. In this regard, flood geology is not only unscientific, it is unbiblical. The first chapter of Romans states that the Creator's divine nature is manifest in His physical creation – in nature. If nature cannot be trusted to tell a truthful story, what does this say about flood geologists' conception of God?

Is Flood Geology an Alternative to Modern Geology?

If the basis for evaluating the plausibility of a suggested history of the Grand Canyon is physical evidence, flood geology falls far short. Where flood geology's explanations appear strong, closer scrutiny invariably finds that critical observations or data that conflict with a global deluge have simply been left out of the discussion. A truly viable alternative must take into account *all* available data, not just the data that fit the model. The flood geology claim of being "as good as" the prevailing scientific view is hollow.

The Grand Canyon provides overwhelming evidence that the Earth is old. Our collective hope is that all who read this book will have the chance to

Figure 20-3. Last light on Comanche Point along the Palisades of the Desert from the Cardenas Creek area. *Photo by Stephen Threlkeld.*

Figure 20-4. A waning flash flood cascades through North Canyon. *Photo by Bronze Black.*

explore the canyon firsthand, to revel in the grandeur of its landscape, and be awed by the incredible story preserved in its layers. Not an imagined history, but the history told by the creation itself.

Does It Really Matter?

Consider for a moment the days of Galileo, when accumulated evidence began to suggest that the Earth was not stationary at the center of the solar system. What would have happened if Galileo and his colleagues had bowed to the accepted wisdom of their society — whether from Ptolemy, the contemporary understanding of select verses from the Bible, or the views of ancient scholars? What if they had simply sought ways to shoehorn their findings into a model in which the Earth was the center of the universe? Assuming all subsequent researchers followed suit, the outcome — the answer to our question — would have been chilling. Every time we follow the directions of our GPS systems, check out the stunning imagery on Google Earth, or watch a TV program via satellite, we are using things that would not exist, had those early scientists failed to allow the data to take them,

Figure 20-5. Profile of the Coconino Sandstone along the Tanner Trail. *Photo by Gregg Davidson.*

unfettered, wherever it led. Launching satellites into space, sending men to the moon, and viewing images from Mars are entirely dependent on the acceptance and understanding of a Sun-centered solar system. Galileo could not have dreamed of a handheld GPS unit, but its very existence traces its lineage directly to his insight and persistence. Science has to be allowed to go where the data leads.

The preceding paragraph touches only on technological advances. What about religious views and sentiments? It is equally important here to allow creation to freely communicate its story. Imagine the effect, during the four centuries since Galileo, if every new celestial observation was forced into an Earth-centered view based on the assumption that biblical verses like Psalm 104:5 ("He set the earth on its foundations, so that it should never be moved"). was intended for instruction on the workings of nature? Believers would understandably become increasingly suspicious not only of natural observations, but more importantly, of God's role in creating and sustaining His creation. Many would eventually feel compelled to leave the faith altogether in the mistaken notion that science and the Bible are hopelessly at odds. Flood geology marches its adherents inexorably down this road. Science, as described in the pages of this book, does not.

Does it matter? It certainly does! Truth *always* matters.

REFERENCES

Where multiple references are provided, each entry is separated by a bullet (•)

General Reading:

Beus, S. S. and Morales, M. (eds.), 2003, *Grand Canyon Geology, 2nd Ed.* Oxford University Press, 432 pp.

Blakey, R. and Ranney, W., 2008, *Ancient Landscapes of the Colorado Plateau.* Grand Canyon Association, 156 pp.

Boggs, S. Jr., 2006, *Principles of Sedimentology and Stratigraphy, 4th Ed.* Pearson Prentice Hall, 662 pp.

Collinson, J., Mountney, N. and Thompson, D., 2006. *Sedimentary Structures, 3rd Ed.* Dunedin Academic Press Ltd., 302 pp.

Davidson, G. R., 2009, *When Faith and Science Collide: A Biblical Approach to Evaluating Evolution and the Age of the Earth.* Malius Press, 290 pp.

Montgomery, D. R., 2013, *The Rocks Don't Lie: A Geologist Investigates Noah's Flood.* W. W. Norton &Company, 320 pp.

Price, L. G., 1999, *An Introduction to Grand Canyon Geology.* Grand Canyon Association, 64 pp.

Ranney, W., 2012, *Carving Grand Canyon: Evidence, Theories, and Mystery, 2nd Ed.* Grand Canyon Association, 190 pp.

Timmons, J. M. and Karlstrom, K. E. (eds.), 2012, *Grand Canyon Geology: Two Billion Years of Earth's History.* The Geological Society of America, Special Paper 489, 156 pp.

Young, D. A. and Stearley, R. F., 2008, *The Bible, Rocks, and Time – Geological Evidence for the Age of the Earth.* InterVarsity Press, 510 pp.

FOREWORD
Page 9

Report on the expedition of Ives and Newberry: Ives, Lt. J. C., 1861, *Report upon the Colorado River of the West.* House Executive Document No. 90, Part 1, 131 pp.

Newberry's statement on the erosive action of water: Newberry, J. S., 1861, *Report upon the Colorado River of the West.* House Executive Document No. 90, Part 3, p. 46.

Page 10

European-schooled geologists couldn't conceive of processes that occurred in Grand Canyon: Ranney, W., 2013, *Geologists Through Time in the Grand Canyon: From Newberry to a New Century.* In: Quartaroli, R. D. (ed.), *A Rendezvous of Grand Canyon Historians; Ideas, Arguments, and First-Person Accounts,* Grand Canyon Historical Society, pp. 113-118. (3rd Grand Canyon History Symposium, January 26-29, 2012, Grand Canyon National Park).

Scientists, many of whom were Christians, became convinced the Earth was old: Montgomery, D. R., 2012, *The Rocks Don't Lie: A Geologist Investigates Noah's Flood.* W. W. Norton, 302 pp., see p. 140. • Young, D. A. and Stearley, R. F., 2008, *The Bible, Rocks, and Time – Geological Evidence for the Age of the Earth.* InterVarsity Press, 512 pp., see p. 101-131.

Chapter 1: Introduction
Page 15

Two bold claims of Young Earth Creationists: Austin, S. A. (ed.), 1994, *Grand Canyon: Monument to Catastrophe.* Institute for Creation Research, 284 pp., see p. 2-4. • Vail, T., 2003, *Grand Canyon: A Different View.* Master Books, 104 pp., see p. 8, 22-23. • Vail, T., Oard, M., Bokovoy, D. and Hergenrather, J., 2008, *Your Guide to the Grand Canyon: A Different Perspective.* Master Books, 190 pp., see p. 141-142, 146-147.

Page 16

Flood geologist's view of how young Earth and old Earth geologists differ: Austin, 1994, *Grand Canyon: Monument to Catastrophe,* p. 21-23.

Chapter 2: What Is Flood Geology?
Page 21

Pairs of every animal: Technically, the animals preserved on the Ark were pairs of all the unclean animals, but seven (or seven pairs) of clean animals.

Page 22

Steno's principles for interpreting rock strata: Young and Stearley, 2008, *The Bible, Rocks, and Time,* p. 53-57.

Two schools of thought among the first geologists: Cutler, A., 2003, *The Seashell on the Mountaintop: How Nicolaus Steno Solved an Ancient Mystery and Created a Science of the Earth.* Plume Books, 228 pp.

Page 23

The Fundamentals allowed for the possibility of an ancient creation: Numbers, R. L., 2006, *The Creationists: From Scientific Creationism to Intelligent Design, Expanded Ed.* Harvard University Press, 606 pp., see p. 53. [In Volume 1, Chapter XIV of *The Fundamentals* (1909), the Rev. Dyson Hague stated: "Genesis is admittedly not a scientific history. It is a narrative for mankind to show that this world was made by God for the habitation of man, and was gradually being fitted for God's children." (URL: user. xmission.com/~fidelis/volume1/chapter14/hague2. php).]

Price sought, but did not receive support from, Christian geologists of the day: Numbers, 2006, *The Creationists: From Scientific Creationism to Intelligent Design, Expanded Ed*, p. 106.

Book that introduced flood geology to modern evangelicals and fundamentalists: Whitcomb, J. C., Jr. and Morris, H. M., 1961, *The Genesis Flood*. Baker Book House, 518 pp.

In *The Creationists*, Numbers documents several discussions between Whitcomb and Morris on how much weight they should give to Price's arguments in *The Genesis Flood*. He states that they ended up "deleting all but a few incidental references to Price and any mention of his Adventist connections", see p. 215-216, 223. In the first edition of *The Genesis Flood*, Price is only cited on four different pages, see p. 184, 185, 189, 211.

Examples of books on the Grand Canyon with origins in The Genesis Flood: Austin, 1994, *Grand Canyon: Monument to Catastrophe*. • Vail, 2003, *Grand Canyon: A Different View*. • Vail, Oard, Bokovoy and Hergenrather, 2008, *Your Guide to the Grand Canyon: A Different Perspective*.

Page 24

According to flood geology, all civilizations discovered by archaeology must fit into the last 4,285 years: Osgood, J., 1981, *The Date of Noah's Flood*. Ex Nihilo, Vol. 4, No. 1, p. 10-13.

Flood geologists have grown increasingly skeptical about the vapor canopy and "no rain before the flood": Morton, G. R., 1979, *Can the Canopy Hold Water?* Creation Research Society Quarterly, Vol. 16, No. 3, p. 164-169. • Snelling, A. A., 2009, *Earth's Catastrophic Past: Geology, Creation, and the Flood*, Institute for Creation Research, 1102 pp., see p. 673. • Mitchel, T., 2010, *There Was No Rain Before the Flood*, Answers in Genesis, URL: answersingenesis.org/ creationism/arguments-to-avoid/there-was-no-rain-before-the-flood • Morris, J. D., 2012, *The Global Flood*. Institute for Creation Research, 175 pp., see p. 34, 55.

Examples of different flood geologist explanations for the "windows of heaven" and "the fountains of the great deep" in Genesis 7:11: Vapor Canopy: Whitcomb and Morris, 1961, *The Genesis Flood*, p. 77. • Catastrophic Plate Tectonics: Austin, S. A., Baumgardner, J. R., Humphreys, D. R., Snelling, A. A., Vardiman, L. and Wise, K. P., 1994, *Catastrophic Plate Tectonics: A*

Global Flood Model of Earth History. In: Walsh, R. E. (ed.), *Proceedings of the Third International Conference on Creationism*, Creation Science Fellowship, p. 609-621, see p. 612. • Hydroplate Theory: Brown, W., 2008, *In the Beginning: Compelling Evidence for Creation and the Flood, 8th Ed*. Center for Scientific Creationism, 448 pp., see p. 105-141. [Brown argues that before the flood, much of Earth's water was stored under the land. At the start of the flood, the land split up into "hydroplates," which then began to sink, causing the water to jet skyward..]

Page 25

According to flood geologists, the fall brought physical death upon Adam and Eve and all subsequent generations of every living creature: Ham, K., 2009, *Was There Death Before Adam Sinned?* In: Ham, K. (ed.), *The New Answers Book 3*, Master Books, 381 pp., see p. 109-117.

Page 26

Scriptural inerrancy as the one basic premise for understanding creation history: Answers in Genesis, *The AiG Statement of Faith* (updated 2012), URL: answersingenesis.org/about/faith. [In section 2 of this statement, we find: "The 66 books of the Bible are the written Word of God. The Bible is divinely inspired and inerrant throughout. Its assertions are factually true in all the original autographs. It is the supreme authority in everything it teaches. Its authority is not limited to spiritual, religious, or redemptive themes but includes its assertions in such fields as history and science." In section 4, we find: "By definition, no apparent, perceived or claimed evidence in any field, including history and chronology, can be valid if it contradicts the scriptural record.".]

How Psalm 104:8 is quoted in *Grand Canyon: A Different View*: Vail, 2003, *Grand Canyon: A Different View*, p 5.

Text of the Chicago Statement of Biblical Inerrancy: International Council on Biblical Inerrancy, 1977, *The Chicago Statement of Biblical Inerrancy*. Dallas Theological Seminary Archives, 8 pp. URL: library. dts.edu/Pages/TL/Special/ICBI_1.pdf.

Page 27

Importance of understanding the ancient Near East mindset: Walton, J. H., 2001, *Genesis NIV Application Commentary*. Zondervan, 759 pp., see p. 25-27, 82-109. • Walton, J. H., 2009, *The Lost World of Genesis One*. InterVarsity Press, 191 pp., see p. 7-13.

Flood geology contradicts the literal biblical location for the Garden of Eden: *The Bible*, Genesis 2:10-14. • Hill, C. A., 2000, *The Garden of Eden: A Modern Landscape*. Perspectives on Science and Christian Faith,

Vol. 52, No. 1, p. 31-46.

Page 28

According to flood geology, people settling the Earth after the flood gave names to geographic features that were the same as names used before the flood: Ham, K., 2009, *Where Was the Garden of Eden Located?* In: *The New Answers Book 3*, p. 15. • Morris, J., 2012, *The Global Flood*, p. 55.

Chapter 3: Time Frame of Flood Geology

Page 31

Flood geologists divide Earth history into five periods: Austin, 1994, *Grand Canyon: Monument to Catastrophe*, p. 57-82. • Vail, 2003, Grand Canyon: *A Different View*, p. 36-37.

Page 33

Flood geologists disagree on which layers are flood deposits: Senter, P., 2011, *The Defeat of Flood Geology by Flood Geology*. Reports of the National Center for Science Education, Vol. 31, No. 3, p. 1.1-1.14.

Page 34

According to flood geologists, metamorphic and igneous basement rock formed during creation week: Austin, 1994, *Grand Canyon: Monument to Catastrophe*, p. 59-60. • Snelling, A. A. and Vail, T., 2009, *When and How Did the Grand Canyon Form?* In: *The New Answers Book 3*, p. 174. • Vail, 2003, *Grand Canyon: A Different View*, p. 36-37.

Page 35

Thickness of the Grand Canyon Supergroup: Hendricks, J. D. and Stevenson, G. M., 2003, *Grand Canyon Supergroup: Unkar Group*. In: Beus, S. S. and Morales, M. (eds.), *Grand Canyon Geology, 2nd Ed.* Oxford University Press, 432 pp., see p. 39-52. • Ford, T. D. and Dehler, C. M., 2003, *Grand Canyon Supergroup: Nankoweap Formation, Chuar Group, and Sixtymile Formation*. In: *Grand Canyon Geology, 2nd Ed.*, p 53-75. • Snelling and Vail, 2009, *When and How Did the Grand Canyon Form?* In: *The New Answers Book 3*, p. 174.

Grand Canyon Supergroup classified as "pre-flood" by flood geologists: Austin, S. A. and Wise, K. P., 1994, *The Pre-Flood/Flood Boundary: As Defined in Grand Canyon, Arizona and Eastern Mojave Desert, California*. In: *Proceedings of the Third International Conference on Creationism*, p. 37-47. • Austin, 1994, *Grand Canyon: Monument to Catastrophe*, p. 62-66. • Wise, K. P. and Snelling, A. A., 2005, *A Note on the Pre-Flood/Flood Boundary in the Grand Canyon*. Origins, No. 58, p. 7-29. • Snelling, A. A., 2008, *Thirty Miles of Dirt in a Day*. Answers Magazine, Vol. 3, No. 4, p. 80-83. [Here, Snelling states that the Grand Canyon Supergroup

and other Precambrian sedimentary layers around the world were laid down in a day – Day 3 of the Creation Week. However, this contradicts the Wise and Snelling (2005) reference listed above, which concludes that the Chuar Group of the Grand Canyon Supergroup was "deposited in a shallow marine to intertidal environment" (p. 26).] • Snelling, 2009, *Earth's Catastrophic Past: Geology, Creation, and the Flood*, p. 709. [Austin, Wise, and Snelling now consider the top formation of the Grand Canyon Supergroup (the highly localized Sixtymile Formation), to be an early flood deposit.]

Start of the early flood period according to flood geology: Snelling, 2009, *Earth's Catastrophic Past: Geology, Creation, and the Flood*, p. 683.

Page 36

Flood onset depicted by flood geologists as giant, planet-circling tsunamis: Video: Answers in Genesis Creation Museum, Petersburg, Kentucky, viewed on November 14, 2014. URL: youtube.com/h?index=68&v= FlPdk4j5R1Y&feature= PlayList&list= PLDF4 B71634466EA3C.

Source of "the fountains of the great deep" according to flood geology: Snelling, 2009, *Earth's Catastrophic Past: Geology, Creation, and the Flood*, p. 697-698.

Alternative flood geology explanation for "fountains of the great deep:" Brown, 2008, *In the Beginning: Compelling Evidence for Creation and the Flood, 8th Ed.*, p. 105-141.

According to flood geologists, Grand Canyon Supergroup tilted at the start of the flood: Austin, 1994, *Grand Canyon: Monument to Catastrophe*, p. 66-67.

Source of sediment according to the first modern flood geologists: Whitcomb and Morris, 1961, *The Genesis Flood*, p. 265, 269.

Source of sediment according to more recent flood geologists: Austin, Baumgardner, Humphreys, Snelling, Vardiman and Wise, 1994, *Catastrophic Plate Tectonics: A Global Flood Model of Earth History*, p. 611.

According to flood geologists, Tapeats Sandstone up to Kaibab Formation are early flood deposits: Austin, 1994, *Grand Canyon: Monument to Catastrophe*, p. 57-58, 67-77. • Vail, 2003, *Grand Canyon: A Different View*, p. 36-37.

According to flood geologists, Grand Staircase is comprised of late flood layers: Austin, 1994, *Grand Canyon: Monument to Catastrophe*, p. 58, 77-79. • Vail, 2003, *Grand Canyon: A Different View*, p. 36-37.

Page 37

Flood geologists' understanding of Genesis 8:3 and 8:5: Austin, 1994, *Grand Canyon: Monument to Catastrophe*,

p. 66-67.

According to flood geologists, receding floodwaters stripped away sediment down to Kaibab Formation: Austin, 1994, *Grand Canyon: Monument to Catastrophe*, p. 78-79.

According to flood geologists, some dinosaurs survived the early flood period: Snelling, 2009, *Earth's Catastrophic Past: Geology, Creation and the Flood*, p. 748-749, 755. • Oard, M. J., 2009, *Dinosaur Tracks, Eggs, and Bonebeds*. In: Oard, M. and Reed, J. K. (eds.), *Rock Solid Answers*, Master Books, 272 pp., see p. 245-258. • Vardiman, L., 1999, *Over the Edge*, Master Books, 160 pp., see p. 17. [This conflicts with the young-earth videos and illustrations depicting giant tsunamis circling the Earth at the start of the flood. How could any dinosaur survive the onslaught of such enormous waves?]

Page 38

Flood geologist's opinions vary on where the flood/post flood boundary is: Austin, Baumgardner, Humphreys, Snelling, Vardiman and Wise, 1994, *Catastrophic Plate Tectonics: A Global Flood Model of Earth History*, p. 614. • Snelling, 2009, *Earth's Catastrophic Past: Geology, Creation and the Flood*, p. 751-761.

According to flood geologists, Claron Formation is the first post-flood layer in the Grand Canyon region: Austin, 1994, *Grand Canyon: Monument to Catastrophe*, p. 58.

Chapter 4: Time Frame of Modern Geology
Page 41

Age of planet Earth: Freedman, R. G. and Kaufmann, W. J., 2010, *Universe: The Solar System, 4th Ed.* W. H. Freeman, 400 pp., see p. 189.

Page 44

There are several places where sediments from every geologic period are found in one thick stack: Morton, G. R., 2001, *The Geologic Column and its Implications for the Flood*. The TalkOrigins Archive, URL: talkorigins.org/faqs/geocolumn • Robertson Group, 1989. *Stratigraphic Database of Major Sedimentary Basins of the World.* • Trendall, A. F. (ed.), 1990, *Geology and Mineral Resources of Western Australia*. Geological Survey of Western Australia, Memoir 3, 827 pp., see p. 382, 396.

Page 45

Some minerals in the schist come from rock that formed up to 3.3 billion years ago: Karlstrom, K. E., Ilg, B. R., Hawkins, D. P., Williams, M. L., Drummond, G., Mahan, K. and Bowring, S. A.,2012, *Vishnu basement rocks of the Upper Granite Gorge: Continent

formation 1.84-1.66 billion years ago. In: Timmons, J. M. and Karlstrom, K. E. (eds.), *Grand Canyon Geology: Two Billion Years of Earth's History,* The Geological Society of America, Special Paper 489, p. 7-24, see p 14-15.

Page 46

Signs left in metamorphic rocks, such as the size, shape, and composition of crystals, are consistent with a history of eventual burial deep enough to begin alteration of the minerals' composition and character: Karlstrom, K. E., Ilg, B. R., Williams, M. L., Hawkins, D. P., Bowring, S. A. and Seaman, S. J., 2003, *Paleoproterozoic Rocks of the Granite Gorges.* In: *Grand Canyon Geology, 2nd Ed.*, p. 26.

Page 47

Fossils in the Unkar Group: Hendricks and Stevenson, 2003, *Grand Canyon Supergroup: Unkar Group.* In: *Grand Canyon Geology, 2nd Ed.*, p. 44, 48. • Timmons, J.M., Bloch, J., Fletcher, K. Karlstrom, K. E., Heizler, M. and Crossey, L. J., 2012, *The Grand Canyon Unkar Group: Mesoproterozoic basin formation in the continental interior during supercontinent assembly.* In: *Grand Canyon Geology: Two Billion Years of Earth's History*, p. 29.

Age of ash layer at base of the Unkar Group (Bass Formation): Timmons, J. M., Karlstrom, K. E., Heizler, M. T., Bowring, S. A., Gehrels, G. E. and Crossey, L. J., 2005, *Tectonic inferences from the ca. 1255-1100 Ma Unkar Group and Nankoweap Formation, Grand Canyon: Intracratonic deformation and basin formation during protracted Grenville orogenesis.* Geological Society of America Bulletin, Vol. 117, No. 11-12, p. 1573-1595.

Fossils in the Chuar Group: Ford and Dehler, 2003, *Grand Canyon Supergroup: Nankoweap Formation, Chuar Group, and Sixtymile Formation.* In: *Grand Canyon Geology, 2nd Ed.*, p. 56, 59, 61-61, 64-70. • Dehler, C. M., Elrick, M., Karlstrom, K. E., Smith, G. A., Crossey, L. J. and Timmons, J. M., 2001, *Neoproterozoic Chuar Group (~800-742 Ma), Grand Canyon: A record of cyclic marine deposition during global cooling and supercontinent rifting.* Sedimentary Geology, Vol. 141, p. 465-499.

Page 48

Over one billion years gap where the Tapeats Sandstone rests on schist: Hendricks and Stevenson, 2003, *Grand Canyon Supergroup: Unkar Group.* In: *Grand Canyon Geology, 2nd Ed.*, p. 43.

Page 49

Fossils in Grand Canyon layers above the Great Unconformity are all Paleozoic: Beus and Morales, 2003, *Introducing the Grand Canyon.* In: *Grand Canyon Geology, 2nd Ed.*, p. 7. • Blakey, R. C. and Middleton, L. T., 2012, *Geologic history and paleogeography of*

Paleozoic and early Mesozoic sedimentary rocks, eastern Grand Canyon. In: *Grand Canyon Geology: Two Billion Years of Earth's History,* p. 82.

In the Grand Staircase, above the older Grand Canyon layers, we find fossils that are absent in the Grand Canyon: dinosaurs, marine reptiles, flying reptiles, flowering plants, and, eventually, early mammals in the upper layers: Doelling, H. H., Blackett, R. E., Hamblin, A. H., Powell, J. D. and Pollock, G. L., *Geology of Grand Staircase-Escalante National Monument, Utah.* In: Sprinkel, D. A., Chidsey, T. C. Jr. and Anderson, P. B. (eds.), 2010, *Geology of Utah's Parks and Monuments, 3rd Ed.* Utah Geological Association, 621 pp., see p. 212-227. • Eaton, J. G., Cifelli, R. L., Hutchinson, J. H., Kirkland, J. I. and Parrish, J. M., (1999), *Cretaceous vertebrate faunas from the Kaiparowits Plateau, south-central Utah.* In: Gillete, D. D. (ed.), *Vertebrate Paleontology in Utah.* Miscellaneous Publication 99-1, Utah Geological Survey, p. 345–353. • DK Publishing, 2012, *Prehistoric Life: The Definitive Visual History of Life on Earth.* DK Adult, 512 pp., see p. 195-357.

Page 50

Lake Claron deposits occurred when Grand Canyon/ Grand Staircase rocks were being uplifted: Hintze, L. F. and Kowallis, B. J., 2009, *Geologic History of Utah.* Dept. of Geological Sciences, Brigham Young University, 225 pp., see p. 79-83.

Chapter 5: Sedimentary Rock Types and How They Form

Page 61

Part of the Claron Formation is known to be a lake deposit: Hintze and Kowallis, 2009, *Geologic History of Utah,* p. 81. • Davis, G. H. and Pollock, G. L., 2010, *Geology of Bryce Canyon National Park, Utah.* In: *Geology of Utah's Parks and Monuments, 3rd Ed.,* p. 46. [Only the White Member of the Claron Formation has been confirmed as being a lake limestone. Jeff Eaton, paleontologist, personal communication, 2013.]

Page 62

Box: Shale makes up roughly 50 percent of sedimentary rock in Earth's geologic record: Boggs, S. Jr., 2006, *Principles of Sedimentology and Stratigraphy, 4th Ed.* Pearson Prentice Hall, 662 pp., see p. 140.

Box: Some flood geologists claim that great stockpiles of sediment (what they refer to as "a substantial thickness of all types of sediment"), totaling over 100 million cubic miles, provided the grains for sedimentary rocks deposited during the flood: Austin, Baumgardner, Humphreys, Snelling, Vardiman and Wise, 1994,

Catastrophic Plate Tectonics: A Global Flood Model of Earth History, p. 611. • Austin, S. A, 2010, *Submarine Liquefied Sediment Gravity Currents: Understanding the Mechanics of the Major Sediment Transportation and Deposition Agent during the Global Flood.* Presentation at 4th Creation Geology Conference, Truett McConnell College, Cleveland, Georgia. [The first sentence in the abstract for this presentation was: "What was the mechanics of the process that transported more than one hundred million cubic miles of sediment during the global flood?"] • Blatt, H., Middleton, G. and Murray, R, 1980. *Origin of Sedimentary Rocks, 2nd Ed.* Prentice-Hall, 782 pp., see p. 34. [Here it states the volume of sedimentary rock accumulated during the Phanerozoic Eon is 654 million cubic kilometers, or 157 million cubic miles. (The Phanerozoic Eon comprises all the so-called early-flood, late-flood, and post-flood layers.).]

Page 63

Box: Popular flood geology explanation for the Tapeats Sandstone, Bright Angel Shale, and Muav Limestone: Austin, 1994, *Grand Canyon: Monument to Catastrophe,* p 67-70.

According to flood geologists, limestones formed when calcite-rich hot water currents encountered cooler currents loaded with lime sediment: Austin, 1994, *Grand Canyon: Monument to Catastrophe,* p. 72.

According to flood geologists, sediment making up thick limestone formations was derived from preexisting stockpiles of lime sediment (what they refer to as "substantial quantities of very fine detrital carbonate sediment") transported from some distance to the canyon: Austin, Baumgardner, Humphreys, Snelling, Vardiman and Wise, 1994, *Catastrophic Plate Tectonics: A Global Flood Model of Earth History,* p. 611.

Page 64

According to flood geologists, thick coral reefs seen around the world could have grown very quickly since the flood: Whitmore, J. H., 2012, *Massive Modern Reefs – Finding Time to Grow.* Answers, Vol. 8, No. 1, p. 72-75. [Whitmore makes an incorrect statement in this reference: "Drilling showed that the mound (Eniwetok Atoll) consists mostly of 'chalky' limestone material, not coral reef organisms." The mound consists of coral reef remains from top to bottom, see: Schlanger, S. O., 1963, *Subsurface geology of Eniwetok Atoll.* USGS Professional Paper 260-BB, p. 991-1066, see p. 1011-1038. URL: pubs.usgs.gov/pp/0260bb/ report.pdf] • Saller, A. H., 1984., *Diagenesis of Cenozoic Limestone on Enewetak Atoll,* PhD Dissertation, Louisiana State University, 362 pp. • Quinn, T. M.

and Saller, A. H., 1997, *Geology of The Anewetak Atoll, Republic of the Marshall Islands.* In: Vacher, H. L. and Quinn, T. (eds.), *Geology and Hydrogeology of Carbonate Islands, Developments in Sedimentology,* Vol. 54, p. 637-665.

According to flood geologists, sediments in the Grand Canyon's layers were transported by hyper-concentrated gravity currents: Austin, S. A., 2012, *Grand Canyon, Creation, and the Global Flood.* Christian Research Journal, Vol. 35, No. 1, p. 50-53. • Snelling, A. A., 2012, *No. 1 Very Little Sediment on the Seafloor,* In: Answers in Genesis, *10 Best Evidences From Science That Confirm a Young Earth.* Answers Magazine, Vol. 7, No. 4, p. 44-58, see p. 47. • Stansbury, D. S., 2013, *How Does an Underwater Debris Flow End?: Flow Transformation Evidences Observed Within the Lower Redwall Limestone of Arizona and Nevada.* In: Horstemeyer, M.218

, Proceedings of the Seventh International Conference on Creationism, Creation Science Fellowship, 45 pp., see p. 1, 13, 26-29, 40-41.

Page 65

Turbidity current deposits take the form of graded beds: Boggs, 2006, *Principles of Sedimentology and Stratigraphy,* p. 359. • Shanmugam, G., 2002, *Ten Turbidite Myths.* Earth-Science Reviews, Vol. 58, p. 311-341, see p. 322-323.

According to flood geologists, Coconino Sandstone was deposited by two to four mile per hour currents: Austin, 1994, *Grand Canyon: Monument to Catastrophe,* p. 33-36. • Austin, 2012, *Grand Canyon, Creation, and the Global Flood,* p. 51. • Snelling, 2009, *Earth's Catastrophic Past: Geology, Creation and the Flood,* p. 507.

What really would be required to deposit the Coconino Sandstone in a matter of days: Helble, T. K., 2011, *Sediment Transport and the Coconino Sandstone: A Reality Check on Flood Geology.* Perspectives on Science and Christian Faith, Vol., 63, No. 1, p. 25-41, see p. 35.

Box: Flood geologists use fish that were buried and preserved in the act of eating another fish as evidence for a global flood: Answers in Genesis, 2007, *Answers Learned from a Fossil.* Answers Magazine, Vol. 2, No. 2, p. 89. • Brown, 2008, *In the Beginning: Compelling Evidence for Creation and the Flood, 8th Ed.,* p. 10.

Chapter 6: Sedimentary Structures: Clues from the Scene of the Crime
Page 69

Raindrop prints have been reported in the Coconino Sandstone: Middleton, L. T., Elliott, D. K. and Morales, M., 2003, *Coconino Sandstone.* In: *Grand Canyon Geology, 2nd Ed.,* p. 171, 173.

Page 70

Maximum angle for loose, dry sand is about 30 to 34°: Lancaster, N., 1995, *The Geomorphology of Desert Dunes.* Routledge, 290 pp., see p. 93-100. • Bagnold, R. A., 1941, *The Physics of Blown Sand and Desert Dunes.* Dover Publications, 265 pp., see p. 201.

Chapter 7: Using the Present to Understand the Past
Page 73

Flood geologists misrepresent uniformitarianism as being synonymous with materialism or evolutionism: Morris, 1961, *The Genesis Flood,* p. 96. • Austin, 1994, *Grand Canyon: Monument to Catastrophe,* p. 22. • Vail, Oard, Bokovoy and Hergenrather, 2008, *Your Guide to the Grand Canyon,* p 15.

According to flood geologists, Mt. St. Helens helps explain rapid formation of the Grand Canyon: Austin, 1994, *Grand Canyon: Monument to Catastrophe,* p. 94-98. • Austin, 2009, *Why is Mount St. Helens Important to the Origins Controversy?* In: *The New Answers Book 3,* p. 253-262, see p. 256-258. • Snelling, 2009, *Earth's Catastrophic Past: Geology, Creation and the Flood,* p. 717-718.

Chapter 8: Solving Puzzles: Relative Dating and the Geologic Column
Page 82

Geologic principles worked out in the 17th century by Nicholas Steno: Cutler, 2003, *The Seashell on the Mountaintop: How Nicolaus Steno solved an Ancient Mystery and Created a Science of the Earth,* 228 pp. • Montgomery, 2012, *The Rocks Don't Lie,* p. 56-64.

Geologic principles contributed in the early nineteenth century by William Smith: Winchester, S., 2001, *The Map That Changed the World.* Harper Collins, 329 pp. • Montgomery, 2012, *The Rocks Don't Lie,* p. 118.

Page 84

Discovery that fossils change regularly in successive layers came far in advance of Darwin's publications: Young and Stearley, 2008, *The Bible, Rocks, and Time,* p. 108.

Page 85

Box: Critics of modern geology say geologists use circular reasoning by dating the fossils by the rocks they are in and dating the rocks by the fossils that are in them: Morris, H. M., 1974, *Scientific Creationism.* Master Books, 281 pp., see p. 229. • Patterson, R., 2008, *Evolution Exposed: Earth Science.* Answers in Genesis, 300 pp., see p. 122.

Page 86

Noah's flood was thought by some to represent the most recent catastrophe: Young and Stearley, 2008, *The Bible, Rocks, and Time*, p. 94.

Page 87

Early descriptions of fossils in the Kaibab Formation of the Colorado Plateau region: Darton, N. H., 1910, *A reconnaissance of parts of northwestern New Mexico and northern Arizona.* U. S. Geological Survey Bulletin 435, 88 pp., see p. 30. • McKee, E. D., 1938, *The environment and history of the Toroweap Formation and Kaibab formations of northern Arizona and southern Utah.* Publication No. 492, Carnegie Institution of Washington, 268 pp. + 48 plates, see p. 153-172. URL: babel.hathitrust.org/cgi/pt?id=mdp.39015078604892;view=1up;seq=171

Permian rocks were first described by Sir Roderick Impey Murchison: Murchison, R. I., De Verneuil, E. and Von Keyserling, A.,1845, *The Geology of Russia in Europe and the Ural Mountains.* John Murray, 826 pp.

Arthur Holmes' first radiometric date for a rock: Lewis, C., 2002, *The Dating Game -One Man's Search for the Age of the Earth.* Cambridge University Press, 272 pp.

Creationists do not question the general validity of the geologic column: Morris, 1974, *Scientific Creationism*, p. 116.

Chapter 9. So Just How Old Is That Rock?
Page 90

Young Earth claim: radiometric dating is unreliable because daughter atoms can be present at the start: DeYoung, D., 2005, *Thousands... Not Billions: Challenging an Icon of Evolution, Questioning the Age of the Earth.* Master Books, 190 pp., see p 42. • Snelling, A. A., 2009, *Radiometric Dating: Problems with the Assumptions.* Answers Magazine, Vol. 4, No. 4, p. 70-73. • Vail, 2003, *Grand Canyon: A Different View*, p. 38.

Scientists now have accurate ways to determine how many daughter atoms were present at the start: Wiens, R. C., 2002, *Radiometric Dating: A Christian Perspective.* URL: asa3.org/ASA/resources/wiens.html • Dickin, A. P., 2005, *Radiogenic Isotope Geology.* Cambridge University Press, 512 pp.

Page 91

Oldest rocks are found in continental "shields," in parts of Canada and Australia: Bowring, S. A. and Williams, I. S., 1999, *Priscoan (4.00-4.03 Ga) orthogneisses from northwestern Canada.* Contributions to Mineralogy and Petrology, Vol. 134, No. 1, p, 3-16.

Page 92

Age for ash bed at the bottom of the Unkar Group: Timmons, Karlstrom, Heizler, Bowring, Gehrels and Crossey, 2005, *Tectonic inferences from the ca. 1255-1100 Ma Unkar Group and Nankoweap Formation, Grand Canyon: Intracratonic deformation and basin formation during protracted Grenville orogenesis*, p. 1573-1595.

Age for Cardenas Lava at the top of the Unkar Group: Timmons, Bloch, Fletcher, Karlstrom, Heizler and Crossey, *The Grand Canyon Unkar Group: Mesoproterozoic basin formation in the continental interior during supercontinent assembly.* In: Timmons, J. M. and Karlstrom, K. E. (eds.), 2012, *Grand Canyon Geology: Two Billion Years of Earth's History.* The Geological Society of America, Special Paper 489, 156 pp., p. 36.

Metamorphic rocks are not useful for dating, unless you are interested in the timing of metamorphic resetting: Mason, R., 1990, *Petrology of the Metamorphic Rocks.* Cambridge University Press, 240 pp., see p. 9-14.

Study comparing changes in iron mineral alignment in the Unkar Group to alignment changes in datable igneous rocks outside the Grand Canyon area: Elston, D. P., 1987, *Chapter 12: Preliminary polar path from Protoerozoic and Paleozoic rocks of the Grand Canyon region, Arizona.* In: Elston, D. P., coordinator, *Geology of Grand Canyon, Northern Arizona (with Colorado River Guides)*, Field Trip Guidebook T115/315, American Geophysical Union, 239 pp., see p. 119-121.

Page 93

Flood geologists claim that "measurable radiocarbon" was found in coal and diamonds: Baumgardner, J. R., 2005, *Chapter 8: 14C Evidence for a Recent Global Flood and a Young Earth.* In: Vardiman, L, Snelling, A. A. and Chaffin, E. F. (eds.), *Radioisotopes and the Age of the Earth: Results of a Young-Earth Creationist Research Initiative.* Institute for Creation Research, 848 pp., see p. 587-625. • DeYoung, 2005, *Thousands... Not Billions*, p. 51-57.

During sample processing, small amounts of radiocarbon are inevitably incorporated from the air in the lab, so that a truly zero reading is never obtained: Isaac, R., 2007, *Assessing the RATE Project.* Perspectives on Science and Christian Faith, Vol. 59, No. 2, p. 143-146, see p. 144-145. • Bertsche, K., 2008, *Intrinsic Radiocarbon?* In: Author Exchange: Isaac, Perspectives on Science and Christian Faith, Vol. 60, No. 1, p. 35-39, see p. 38. An expanded version of this response is on the American Scientific Affiliation website: *Radiocarbon: Intrinsic or Contamination?* URL: asa3.org/ASA/education/origins/carbon-kb.htm.

The Mars rover is powered by an isotope of plutonium that has a half-life of 88 years: Grotzinger, J. P. et al., 2012, *Mars Science Laboratory mission and science* investigation. Space Science Reviews, Vol. 170, No. 1-4, p. 5-56. • Balint, T. S. and Jordan, J. F., 2007, *RPS strategies to enable NASA's next decade robotic Mars missions.* Acta Astronautica, Vol. 60, p. 992-1001.

Rocks produced by eruption of Mt. Vesuvius have been correctly dated using the argon-argon method to within a few years of the event: Renne, P. R., Sharp, W. D., Deino, A. L., Orsi, G. and Civetta, L., 1997, *40Ar/39Ar dating into the historical realm: calibration against Pliny the Younger.* Science, Vol. 277, No. 5330, p 1279-1280. • Dalrymple, G. B., 2000, *Radiometric Dating Does Work!* Reports of the National Center for Science Education, Vol. 20, No. 3, p. 14-19. [This paper includes four case studies where radiometric dating resulted in very similar results. For example, for the time of the extinction of the dinosaurs, the results of 187 analyses all gave ages between 63.7 and 66.0 million years.]

Page 94

Spreading rates for the Atlantic Ocean crust range from 1.1 to 1.7 inches/year: Davidson, G. R., 2009, *When Faith and Science Collide.* Malius Press, 288 pp., see p.116-119.

Page 95

Flood geologist obtained an age of 350,000 years for a sample from the fresh cone in Mt. St. Helens: Austin, S. A., 1996, *Excess Argon within Mineral Concentrates from the New Dacite Lava Dome at Mount St. Helens Volcano.* Creation ex nihilo Technical Journal, Vol. 10, No. 3, pp. 335-343.

Potassium-argon dating has long been recognized to yield inaccurate results for recent lava flows: Kelley, S., 2002, *Excess argon in K-Ar and Ar-Ar geochronology.* Chemical Geology, Vol. 188, No. 1-2, p. 1-22. • Dalrymple, G. B. and Moore, J. G., 1968, *Argon-40: Excess in Submarine Pillow Basalts from Kilauea Volcano, Hawaii.* Science, Vol. 161, No. 3846, p. 1132-1135.

Page 96

Flood geologists submitted samples from an igneous intrusion (sill) within the Hakatai Shale for dating using four different methods: Snelling, A. A., Austin, S. A. and Hoesch, W. A., 2003, *Radioisotopes in the diabase sill (upper Precambrian) at Bass Rapids, Grand Canyon, Arizona: An application and test of the isochron dating method.* In: Ivey, R. L., Jr. (ed.), *Proceedings of the Fifth International Conference on Creationism,* Creation Science Fellowship, 597 pp., see p. 269-284.

Page 97

Flood geologist attempted to date recent lava flows in western Grand Canyon: Austin, 1994, *Grand Canyon: Monument to Catastrophe,* p. 123-126.

Chapter 10. Missing Time: Gaps in the Rock Record
Page 100

Sources for information in Figure 10-1 on Grand Canyon unconformities: Hendricks and Stevenson, 2003, *Grand Canyon Supergroup: Unkar Group.* In: *Grand Canyon Geology, 2nd Ed.,* p. 44, 45, 48-49. • Ford and Dehler, 2003, *Grand Canyon Supergroup: Nankoweap Formation, Chuar Group, and Sixtymile Formation.* In: *Grand Canyon Geology, 2nd Ed.,* p. 55, 59. • Timmons, J. M. and Karlstrom, K.E., 2012, *Many unconformities make one 'Great Unconformity.'* In: *Grand Canyon Geology: Two Billion Years of Earth's History,* p. 73-79.

Page 101

The Great Unconformity is one of the few features acknowledged by flood geologists to be a genuine unconformity: Austin, 1994, *Grand Canyon: Monument to Catastrophe,* p. 45-47, 57, 67. • Snelling, 2009, *Earth's Catastrophic Past: Geology, Creation and the Flood,* p. 709.

Page 102

The rock fragments in the Surprise Canyon Formation are clearly from the underlying Redwall Limestone: Billingsley, G. H. and Beus, S. S., 1999, *Chapter D: Erosional Surface of the Surprise Canyon Formation.* In: Billingsley, G. H. and Beus, S. S., *Geology of the Surprise Canyon Formation of the Grand Canyon, Arizona.* Museum of Northern Arizona Bulletin 61, 254 pp., see p. 53-68.

Page 103

The bottom layers of the Surprise Canyon Formation contain fossil plant material and the middle and upper layers contain marine fossils: Billingsley, G. H. and Beus, S. S., 1999, *Chapter E: Megafossil paleontology of the Surprise Canyon Formation.* In: *Geology of the Surprise Canyon Formation of the Grand Canyon, Arizona,* p. 69-96.

Page 105

These features host the highest uranium-ore deposits in North America: Mathisen, I. W., 1987, *Arizona Strip breccia pipe program: Exploration, development, and production (abstract).* American Association of Petroleum Geologists Bulletin, Vol. 71, no. 5, p. 590.

Modern cave passages have exposed much older caves that were filled in long ago after their roofs collapsed: Davis, D. G. and Huntoon, P. W., 2004, *Grand Canyon, United States.* In: Gunn, J. (ed.), *Encyclopedia of Caves and Karst Science,* Fitzroy, 960 pp, see p. 391–393.

Chapter 11. Plate Tectonics: Our Restless Earth
Page 109

Flood geologists claim that the continents spread

apart very rapidly during Noah's flood: Austin, Baumgardner, Humphreys, Snelling, Vardiman and Wise, 1994, *Catastrophic Plate Tectonics: A Global Flood Model of Earth History*, p. 612.

If we examine the driving forces for plate tectonics, we are left with the simple concepts of gravity, heat, and density: McConnell, D., Steer, D., Owens, K. and Knight, C, 2010, *The Good Earth: Introduction to Earth Science, 2nd Ed.* McGraw-Hill, 529 pp., see p. 75-103. • Marshak, S., 2011, *Earth: Portrait of a Planet, 4th Ed.*, W.W. Norton & Company, 819 pp., see p. 77-101. [Several excellent YouTube videos and a very good PowerPoint file are available at www.houstonisd. org, search "plate tectonics."] • Gonzales, G. and Richards, J. W., 2004, *The Privileged Planet – How Our Place in the Cosmos Is Designed for Discovery,* Regnery Publishing, 464 pp., see p. 57-62. [Chapters 2 and 3 explain how the occurrence of radioactive atoms of uranium, thorium, and potassium in our Earth is necessary for plate tectonics and a magnetic field, both of which are critical for life to exist.]

Page 111

Flood geologists say that current movement along plate boundaries is the residual movement as plates have gradually slowed since the days of the flood: Austin, Baumgardner, Humphreys, Snelling, Vardiman and Wise, 1994, *Catastrophic Plate Tectonics: A Global Flood Model of Earth History,* p. 615.

Box: Some flood geologists argue that massive limestone and salt deposits formed when hot, mineral-rich waters were released from the "fountains of the deep:" Brown, 2008, *In the Beginning: Compelling Evidence for Creation and the Flood, 8th Ed.,* p. 122, 221-225.

Box: What we actually find being emitted by black smokers: Von Damm, K. L., 1990, *Seafloor Hydrothermal Activity: Black Smoker Chemistry and Chimneys,* Annual Review of Earth and Planetary Sciences, Vol. 18, p. 173-204.

Page 112

The magnitude of the earthquake that caused the Japan tsunami of 2011 was 9.0, the length of the rupture zone was 155 miles, and the maximum movement at the fault zone was about 88 feet. When averaged since the last rupture, the rate of movement of the Pacific plate smashing into Japan is about 3 inches per year: Lay, R., Fufii, Y., Geist, E., Koketsu, K., Rubinstein, J., Sauya, T. and Simons, M., 2013, *Special Issue on the 2011 Tohoku Earthquake and Tsunami.* Bulletin of the Seismological Society of America, Vol. 103, No. 2B, p. 1165-1627. • Perfettini, H. and Avouac, J.P., 2014, *The Seismic Cycle in the Area of the 2011 Mw 9.0 Tohoku-*

Oki Earthquake. Journal of Geophysical Research: Solid Earth, Vol. 119, No. 5, p. 4469-4515.

Map of tectonic plate boundaries and earthquakes over a certain magnitude over a 10-year period (Fig. 11-5): A similar map is available at the U. S. Geological Survey URL: earthquake.usgs.gov/earthquakes/world/ seismicity_maps.

Page 114

Box: Some flood geologists propose that water blasted heavenward through cracks at supersonic speeds at the start of Noah's flood: Brown, 2008, *In the Beginning: Compelling Evidence for Creation and the Flood, 8th Ed.,* p. 105-141.

Page 115

Sedimentary rock on top of Mt. Everest contains marine fossils (Fig 11-9): Searle, M. P., Simpson, R. L., Law, R. D., Parrish, R. R. and Waters, D. J., 2003, *The structural geometry, metamorphic and magmatic evolution of the Everest massif, High Himalaya of Nepal—South Tibet.* Journal of the Geological Society, London, Vol. 160, p. 345-366, see p. 354.

Chapter 12. Broken and Bent Rock: Fractures, Faults, and Folds

Pages 118

Recognizing the types of faults can speak to us about the type of tectonic forces at work at the time and whether layers were soft sediment or hardened rock at the time of deformation: Marshak, S., 2011, *Earth: Portrait of a Planet, 4th Ed.* W.W. Norton & Company, 819 pp., see p. 348-367.

Page 123

The Bright Angel Fault has had considerable strike slip movement during Precambrian time: Shoemaker, E. M., Squires, R. L. and Abrams, M. J., 1975, *The Bright Angel, Mesa Butte, and related fault systems of Northern Arizona.* In: Goetz, A., Billingsley, F. C., Gillespie, R. R., Abrams, J. J. and Squires, R. L., 1975, *Application of ERTS images and image processing to regional geologic problems and geologic mapping in Northern Arizona,* NASA Technical Report 32-1597, 188 pp., see p. 23-41.

Sequential history of faulting, folding, deposition, and erosion in the Grand Canyon area: Huntoon, P. W., 2003, *Post Precambrian Tectonism in the Grand Canyon Region.* In: *Grand Canyon Geology, 2nd Ed.,* p. 222-259.

Page 125

Flood geologists argue that the sediments of the Tapeats Sandstone were still soft when the Carbon Canyon fold was formed: Snelling, 2009, *Earth's Catastrophic Past: Geology, Creation and the Flood,* p. 599-601. •

Snelling, A. A., 2009, *What Are Some of the Best Flood Evidences?* In: *The New Answers Book 3,* p. 296-298. • Vail, 2003, *Grand Canyon: A Different View,* p. 22-23, 32-33. [Perhaps the earliest flood geologist to use the "folded while still soft" argument (not specifically for the Tapeats Sandstone) was Price, G. M., 1923, *The New Geology,* Pacific Press Publishing Association, 726 pp., see p. 691.]

Page 126

We find both flexural slippage and abundant fractures in the fold: Niglio, L., 2004, *Fracture Analysis of Precambrian and Paleozoic rocks in selected areas of the Grand Canyon National Park, USA.* University of Oklahoma, Norman, Masters Thesis. 68 p. • Sassi, W., Guiton, M. L. E., Leroy, Y. M., Daniel, J. M. and Callot, J. P., 2012, *Constraints on bed scale fracture chronology with a FEM mechanical model of folding; the case of Split Mountain (Utah, USA).* Tectonophysics, Vol. 576-577, p 197-215. [A classic textbook on this subject is: Ramsay, J. G., 1967, *Folding and Fracturing of Rocks.* McGraw-Hill, 568 pp.]

Chapter 13. Fossils of the Grand Canyon and Grand Staircase

Page 131

At least four Young Earth/flood geology explanations for the order in the fossil record: (1) Hydraulic sorting by size or shape: Whitcomb and Morris, 1961, *The Genesis Flood,* p. 266-277. • Morris, H., 1970, *Sedimentation and the Fossil Record: A Study in Hydraulic Engineering.* In: Lammerts, W. (ed.), *Why Not Creation,* Baker Book House, 388 pp. • (2) Catastrophic transport and burial of entire ecosystems: Whitcomb and Morris, 1961, *The Genesis Flood,* p. 276. • Roth, A. A., 1998, *Origins: Linking Science and Scripture,* Review and Herald Publishing, 384 pp., see p. 170-174. • (3) More complex organisms could flee to higher ground: Whitcomb and Morris, 1961, *The Genesis Flood,* p. 266. • (4) Sorting by earthquake vibrations based on size or density: Brown, 2008, *In the Beginning: Compelling Evidence for Creation and the Flood, 8th Ed.,* p. 169-178. [In *The Bible, Rocks and Time,* Young and Stearley examine many actual fossil assemblages to demonstrate that they do not match any of the expectations for these Young Earth proposals, see p. 243-287.]

Page 134

Box: Young Earth advocates say the Cambrian explosion is clear evidence that complex life started all at once: Gish, D. T., 1985, *Evolution: The Challenge of the Fossil Record.* Master Books, 277 pp. • Austin, *Grand Canyon: Monument to Catastrophe,* p. 147. • Wise, K., 2009, *One: Life's Unexpected Explosion: What Explains the Cambrian Explosion?* Answers Magazine, Vol. 5, No. 1, p. 40-41. • Vail, 2003, *Grand Canyon: A Different View,* p. 50-51.

There are many types of complicated organisms now documented as fossils from the uppermost Precambrian strata, AND in some cases these fossils include forms previously considered Cambrian. For more information, see: Bengtson, S. (ed.), 1994, *Early Life on Earth.* Columbia University Press, 656 pp. • Zhuravlev, A. Y. and Riding, R. (eds.), 2001, *The Ecology of the Cambrian Radiation.* Columbia University Press, 576 pp. • Fedonkin, M. A., Gehling, J. G., Grey, K., Narbonne, G. M. and Vickers-Rich, P., 2007, *The Rise of Animals: Evolution and Diversification of the Kingdom Animalia.* Johns Hopkins University Press, 344 pp. • Erwin, D. H. and Valentine, J. W., 2013, *The Cambrian Explosion: The Construction of Animal Biodiversity.* Roberts and Company, 416 pp.

The following three references provide analyses of the Precambrian-Cambrian transition from an evangelical, old-earth perspective: Campbell, D. and Miller, K. B., 2003, *The "Cambrian Explosion": A Challenge to Evolutionary Theory?* In: Miller, K. B. (ed.), *Perspectives on an Evolving Creation,* Wm. B. Eerdmans, 528 pp., see p. 182-207. • Stearley, R., 2013, *The Cambrian Explosion: How Much Bang for the Buck?* Perspectives on Science and Christian Faith, Vol. 65, No. 4, p. 245-257. • Miller, K. B., 2014, *The Fossil Record of the Cambrian "Explosion": Resolving the Tree of Life,* Perspectives on Science and Christian Faith, Vol. 66, No. 2, p. 67-82.

Page 135

Coral or snail fossils in the Grand Canyon are not just different from modern varieties at the species or genus level, they are so distinct that biologists place them in different orders. To better understand the large biological differences between the groups of dead organisms preserved as fossils in the Paleozoic strata of the Grand Canyon vs. the groups of live organisms present nearby today, compare the modern orders and families of corals, brachiopods, bryozoans, arthropods, clams, cephalopods and others found in the two groups of references below. Modern organisms: Brusca, R. C., 1980, *Common Intertidal Invertebrates of the Gulf of California,* 2nd Ed. University of Arizona Press, 513 pp. • Ricketts, E. F., Calvin, J. and Hegpeth, J. W., 1985, Between Pacific Tides, 5th Ed., revised by Phillips, D. W., Stanford University Press, 680 pp. Fossil invertebrates in the Grand

Canyon: Sadler, C., 2007, *Life in Stone: Fossils of the Colorado Plateau*. Grand Canyon Association, 72 pp. • and many references cited in Beus and Morales, Grand Canyon Geology.

Page 136

Box: Leading Young Earth creationists now argue that differences between organisms today and those found as fossils in flood layers, can be explained by adaptation since the time of the flood (from a single pair of progenitors): Ham, K., 2003, *Did God Create Poodles? Creation Magazine,* Vol. 25, No. 4, p. 19-22. • Christian, M., 2007, *Purring Cats and Roaring Tigers.* Answers Magazine, Vol. 2, No. 4, p. 20-22.

Flood geology authors in *Grand Canyon: A Different View* suspect Grand Canyon Supergroup was tilted during early phase of the flood catastrophe: Vail, 2003, *Grand Canyon: A Different View*, p. 36, 46.

Page 142

Flood geologists claim to see evidence of catastrophic transport and also sorting of fossils by various hydrodynamic processes: See references for p. 131.

Box: Flood geologists frequently cite the orientation of nautiloid shells as evidence of a flood of catastrophic proportions: Austin, 1994, *Grand Canyon: Monument to Catastrophe*, p. 26-28. • Snelling, 2009, *Earth's Catastrophic Past: Geology, Creation and the Flood*, p. 497-498. • Vail, 2003, *Grand Canyon: A Different View*, p. 52-53.

Chapter 14. Tiny Plants – Big Impact: Pollen, Spores, and Plant Fossils

Page 148

To find multicellular plants, we have to move up into the overlying Paleozoic layers, where we also find a distinct order: Taylor, W. A. and Strother, P. K., 2008, *Ultrastructure of some Cambrian Palynomorphs from the Bright Angel Shale, Arizona, USA*. Review of Palaeobotany and Palynology, Vol. 151, p. 41-50. • Kenrick, P. and Crane, P. R., 1997, *The Origin and Early Diversification of Land Plants: A Cladistic Study.* Smithsonian Institution, 441 pp.

Page 149

Typical swampland forest in the Permian (Fig 14-3): Wang, J., Pfefferkorn, H. W., Zhang, Y. and Feng, Z, 2012, *Permian Vegetational Pompeii from Inner Mongolia and its Implications for Landscape Paleoecology and Paleobiogeography of Cathaysia.* Proceedings National Academy of Sciences, Vol. 109, No. 13, p. 4927-4932.

The external appearance of these reproductive products is highly variable for land plants: Duff, R. J., 2009, *Flood Geology's Abominable Mystery.* Perspectives on

Science and Christian Faith, Vol. 60, No. 3, p. 162-171. • Tidwell, W. D., Jennings, J. R. and Bues, S. S., 1992, *A Carboniferous Flora from the Surprise Canyon Formation in the Grand Canyon, Arizona.* Journal of Paleontology, Vol. 66, No. 6, p. 1013-1021. • Fleming, R. F., 1994, *Cretaceous Pollen in Pliocene Rocks: Implications for Pliocene Climate in the Southwestern United States.* Geology, Vol. 22, p. 787-790.

Page 150

Cambrian and Devonian rocks, represented by the Tapeats Sandstone up to the Temple Butte Formation, contain only simple trilete (3-lobed) spores: Taylor and Strother, 2008, *Ultrastructure of some Cambrian Palynomorphs from the Bright Angel Shale, Arizona, USA*, p. 41-50.

These spores are associated with the plant remains of ferns, horsetails, and giant lycophyte trees: Tidwell, Jennings and Bues, 1992, *A Carboniferous Flora from the Surprise Canyon Formation in the Grand Canyon, Arizona*, p. 1013-1021. • Fleming, 1994, *Cretaceous Pollen in Pliocene Rocks: Implications for Pliocene Climate in the Southwestern United States*, p. 787-790.

Conifers are the dominant plants in Triassic and Jurassic rocks of the Grand Staircase and their pollen is readily distinguishable from flowering-plant pollen: Fleming, R. F., 1994, *Cretaceous Pollen in Pliocene Rocks: Implications for Pliocene Climate in the Southwestern United States*, Geology, Vol. 22, No. 9, p. 787-790.

Light-microscope images of fossil spores extracted from the Grand Staircase (Fig 14-5): Cretaceous Dakota Formation, Canyonlands area, southern Utah: Am Ende, B. A., 1991, *Depositional environments, palynology, and age of the Dakota Formation, south-central Utah.* In: Nations, J. D. and Eaton, J. G. (eds.), *Stratigraphy, Depositional Environments, and Sedimentary Tectonics of the Western Margin, Cretaceous Western Interior Seaway,* The Geological Society of America Special Paper 260, p. 65-83.

Evidence of the existence of flowering plants in the Grand Canyon is found only in the sediments of modern caves and canyons: Van Devender, T. R. and Mead, J. I., 1976, *Late Pleistocene and Modern Plant Communities of Shinumo Creek and Peach Springs Wash, Lower Grand Canyon, Arizona.* Journal of the Arizona Academy of Science, Vol. 18, p. 16-22. • Cole, K., 1985, *Past Rates of Change, Species Richness, and a Model of Vegetational Inertia in Grand Canyon, Arizona.* The American Naturalist, Vol.125, No. 2, p. 289-303. • Mead, J. I. and Phillips, A. M., 1981, *The Late Pleistocene and Holocene Fauna and Flora of Vulture Cave, Grand Canyon, Arizona.* The Southwestern Naturalist,

Vol. 26, No. 3, p. 257-288.

Page 151

Box: Some flood geologists have claimed to have found pollen from flowering plants in the Hakatai Shale: Chadwick, A. W., 1981, *Precambrian Pollen in the Grand Canyon – A Reexamination.* Origins, Vol. 8, p. 7-12. • Howe, G. F., 1986, *Creation Research Society Studies on Precambrian Pollen: Part I – A Review.* Creation Research Society Quarterly, Vol. 23, No. 3, p. 99-104. • Lammerts, W. E. and Howe, G. F., 1987, *Creation Research Society Studies on Precambrian Pollen – Part II: Experiments on Atmospheric Pollen Contamination of Microscope Slides.* Creation Research Society Quarterly, Vol. 23, No. 4, p. 151-153. • Howe, G. F., Lammerts, E. L., Matzko, G. T., and Lammerts, W. E., 1988, *Creation Research Society Studies on Precambrian Pollen, Part III: A Pollen Analysis of Hakatai Shale and Other Grand Canyon Rocks.* Creation Research Society Quarterly, Vol. 24, No. 4, p. 173-182.

Chapter 15 Trace Fossils: Footprints and Imprints of Past Life

Page 153

Trace fossils are tracks, trails, burrows, borings and other structures made by ancient organisms and then preserved in the fossil record: Prothero, D. R., 2004, *Bringing Fossils to Life: An Introduction to Paleobiology.* McGraw Hill, 503 pp.

Page 154

What is known about trace fossils in the Coconino Sandstone: Middleton, Elliott and Morales, 2003, *Coconino Sandstone.* In: *Grand Canyon Geology, 2nd Ed.,* p. 163-179.

Trace fossils in the Cambrian Bright Angel Shale: Middleton, L. T. and D. K. Elliott, 2003, *Tonto Group.* In: *Grand Canyon Geology, 2nd Ed.,* p. 90-114. • Elliott, D. K. and Martin, D. L., 1987, *A new trace fossil from the Cambrian Bright Angel Shale, Grand Canyon, Arizona.* Journal of Paleontology, Vol. 61, p. 641-648.

Trace fossils in the Supai Group and Hermit Formation: Hunt, A. P., Lucas, S. G., Santucci, V. L. and Elliott, D. K., 2005, *Permian Vertebrates of Arizona.* In: Heckert, A. B. and Lucas, S. G. (eds.), *Vertebrate Paleontology in Arizona,* New Mexico Museum of Natural History and Science Bulletin, Vol. 29, p. 10-15.

The Coconino Sandstone is evidence of an enormous desert sand sea: McKee, E. D., 1933, *The Coconino Sandstone. Its History and Origin.* Carnegie Institute Publication No. 440, p. 77-115.

Traces of animals and their behavior preserved as footprints and burrows in Coconino sands: Gilmore,

C. W., 1928, *Fossil Footprints in the Grand Canyon of the Colorado, Arizona.* Smithsonian Institution Publication 2957, p. 7-10. • Hunt, A. P. and Santucci, V. L., 1998, *Taxonomy and Ichnofacies of Permian Tetrapod Tracks From Grand Canyon National Park, Arizona.* In: Santucci, V. L. and McClelland, L. (eds.), *National Park Service Paleontological Research,* National Park Service Geological Resources Division Technical Report NPS/NRGRD/GRDTR-98, p. 94-96.

Page 157

The footprint area and depth is proportional to the weight of the animal; tracks in the Coconino indicate that the largest animals weighed between 20 and 30 pounds: Alexander, R. M., 1989, *Dynamics of dinosaurs and other extinct giants.* Columbia University Press, p. 31, 32.

A deposit of the same age in Texas contains similar footprints associated with the bones of early synapsids: Lucas, S. G. and Hunt, A. P., 2006, *Permian tetrapod footprints: Biostratigraphy and Biochronology.* Geological Society of London Special Publication 265, p. 179-200, see p. 185.

Experiments with living organisms have shown that these fossil tracks are most similar to modern tracks made by spiders and scorpions: Middleton, Elliott and Morales, 2003, *Coconino Sandstone.* In: *Grand Canyon Geology, 2nd Ed.,* p. 169. • Sadler, C. J. 1993, *Arthropod Trace Fossils from the Permian De Chelly Sandstone, Northeastern Arizona.* Journal of Paleontology, Vol. 67, p. 240-249. • McKee, E. D., 1947, *Experiments on the Development of Tracks in Fine Cross-Bedded Sand.* Journal of Sedimentary Petrology, Vol. 17, p. 23-28.

Page 158

How preservation of tracks can occur after they have been made: Middleton, Elliott and Morales, 2003, *Coconino Sandstone.* In: *Grand Canyon Geology, 2nd Ed.,* p. 169, 170.

Flood geologists claim that vertebrate tracks in the Coconino Sandstone were made by animals attempting to escape advancing flood waters: Brand, L. R., 1979, *Field and laboratory studies on the Coconino Sandstone vertebrate footprints and their paleoecological implications.* Palaeogeography, Palaeoclimatology, and Palaeoecology, Vol. 28, p. 25-38. • Austin, 1994, *Grand Canyon: Monument to Catastrophe,* p. 31-32, 146. [This conflicts with other flood geology claims, since the Coconino Sandstone, along with multiple underlying layers, is said to have been deposited by the flood. If the animals were able to survive for months by swimming or holding onto floating material until the Coconino was deposited towards the end of the early flood period, why would they need to run away from advancing flood waters?]

Page 159

"I wouldn't have seen it if I hadn't believed it": Vail, 2003, *Grand Canyon: A Different View*, p. 42. Here, John Morris applied this maxim to "those who advocate the desert interpretation".

Chapter 16. Carving of the Grand Canyon: A Lot of Time and a Little Water, or a Lot of Water and a Little Time (or Something Else?)

Page 163

Young Earth advocates typically limit the choices to two options: a lot of time and a little water, or a lot of water and a little time: Snelling and Vail, 2009, *When and How Did the Grand Canyon Form?* In: *The New Answers Book 3*, p. 185. • Morris, J., 2002, *A Canyon in Six Days!* Creation, Vol. 24, No. 4, p.54-55.

Page 164

The most prominent flood geologists propose that the Kaibab arch created natural dams that formed large lakes upstream from the present-day Grand Canyon: Austin, 1994, *Grand Canyon: Monument to Catastrophe*, p. 92-104. • Brown, 2008, *In the Beginning: Compelling Evidence for Creation and the Flood, 8th Ed.*, p. 182-219. [On pp. 213-215, Brown states "In early 1990, Austin published, as if they were his, some key ideas of mine concerning Grand (Canyonlands) Lake and the formation of the Grand Canyon."] • Snelling, 2009, *Earth's Catastrophic Past: Geology, Creation and the Flood*, p. 768.

Statement by Henry Morris: "A great dammed-up lake full of water from the Flood suddenly broke and a mighty hydraulic monster roared toward the sea, digging deeply into the path it had chosen along the way.": Vail, 2003, *Grand Canyon: A Different View*, p 4.

This idea (the spillover model) became popular among many geologists at the 2000 Symposium on the Origin of the Colorado River: Young, R. A. and Spamer, E. E. (eds.), 2004, *Colorado River: Origin and Evolution: Proceedings of a Symposium Held at Grand Canyon National Park in June 2000*. Grand Canyon Association, 280 pp.

The spillover model and the breached-dam hypothesis are both now rejected by most geologists because the bulk of the evidence does not support either theory: Dickinson, W. R., 2013, *Rejection of the lake spillover model for initial incision of the Grand Canyon, and discussion of alternatives*. Geosphere, Vol. 9, No. 1, p. 1-20.

Page 165

Recent scientific studies of Bidahochi sediments show that the lake pictured as Hopi Lake by flood geologists

was probably never one big lake: Dallegge, T. A., Ort, M. H. and McIntosh, W. C., 2003, *Mio-Pliocene chronostratigraphy, basin morphology, and paleodrainage relations derived from the Bidahochi Formation, Hopi and Navajo Nations, northeastern Arizona*. The Mountain Geologist, Vol. 40, No. 3, p. 55-82.

Page 166

According to flood geologists, rushing waters formed the main canyon, side canyons formed as soft sediments collapsed into the main canyon and were washed away: Austin, 1994, *Grand Canyon: Monument to Catastrophe*, p. 99-102.

Page 168

Box: Some flood geologists argue for the (post-flood) breached-dam hypothesis, while others argue the Grand Canyon was carved by receding floodwaters from Noah's flood: Breached dam: Austin, 1994, *Grand Canyon: Monument to Catastrophe*, p. 92-107. • Snelling and Vail, 2009, *When and How Did the Grand Canyon Form?* In: *The New Answers Book 3*, p. 183-184 • Brown, 2008, *In the Beginning: Compelling Evidence for Creation and the Flood, 8th Ed.*, p. 188-190. • Late flood: Oard, M. J., 2010, *The Origin of Grand Canyon Part II: Fatal Problems with the Dam-Breach Hypothesis*. Creation Science Research Quarterly, Vol. 46, No. 4, p. 290-307. • Oard, M. J., 2010, *The Origin of Grand Canyon Part III: A Geomorphological Problem*. Creation Science Research Quarterly, Vol. 47, No. 1, p. 45-57. • Oard, M. J., 2010, *The Origin of Grand Canyon Part IV: The Great Denudation*. Creation Science Research Quarterly, Vol. 47, No. 2, p. 146-157.

Question asked by flood geologists: "Where is all the sediment that supposedly was steadily eroded from the canyon over millions of years?": Vail, 2003, *Grand Canyon: A Different View*, p. 30-31. • Austin, 1994, *Grand Canyon: Monument to Catastrophe*, p. 87.

Numerous papers have documented where sediment eroded from the Grand Canyon is: Dorsey, R. J., Housen, B. A., Janecke, S. U., Fanning, C. M. and Spears, A. L., 2011, *Stratigraphic record of basin development within the San Andreas fault system: Late Cenozoic Fish Creek-Vallecito basin, Southern California*. Geological Society of America Bulletin, Vol. 123, No. 5/6, p. 771-793. • Spencer, J.E, Patchett, P. J., Roskowski, J. A., Pearthree, P. A., Faulds, J. E. and House, P. K., 2011, *A Brief Review of Sr Isotopic Evidence for the Setting and Evolution of the Miocene - Pliocene Hualapai-Bouse Lake System*. In: Beard, L. S., Karlstrom, K. E., Young, R. A. and Billingsley, G. H., CRevolution 2—Origin and Evolution of the Colorado River System, Workshop Abstracts. U.S. Geological Survey

Open File Report 2011–1210, 300 pp., see p. 250-259. URL: pubs.usgs.gov/of/2011/1210/of2011-1210.pdf.

Page 169

Four examples of small canyons cited by flood geologists as evidence for rapid carving of the Grand Canyon: Channeled Scablands: Austin, 1994, *Grand Canyon: Monument to Catastrophe*, p. 94, 96-97. • Oard, M., 2008, *Flood By Design,* Master Books, 130 pp., see p. 100-101. • Snelling and Vail, 2009, *When and How Did the Grand Canyon Form?* In: *The New Answers Book 3,* p. 181. • Mt. St. Helens: Austin, 1994, *Grand Canyon: Monument to Catastrophe*, p. 94, 97-98. • Austin, 2009, *Why is Mount St. Helens Important to the Origins Controversy?* In: *The New Answers Book 3*, p. 253-262, see p. 256-258. • Snelling and Vail, 2009, *When and How Did the Grand Canyon Form?* In: *The New Answers Book 3*, p. 181-182. • Snelling, 2009, *Earth's Catastrophic Past: Geology, Creation and the Flood*, p. 718. • Canyon Lake Gorge: Doyle, S., 2009, *A Gorge in Three Days!* Creation Magazine, Vol. 31, No. 3, p. 32-33. • Providence Canyon: Williams, E. L., 1995, *Providence Canyon, Stewart County, Georgia - Evidence of recent rapid erosion.* Creation Research Society Quarterly, Vol. 32, No. 1, p. 29-43. • Gibson, R., 2000, *Canyon Creation: Faster Than Most People Would Think Possible, Beauty Was Born from Devastation.* Creation Magazine, Vol. 22, No. 4, p. 46-48.

Page 171

Volume of flood water and eroded rock for Canyon Lake Gorge event: Lamb, M. P. and Fonstad, M. A., 2010. *Rapid formation of a modern bedrock canyon by a single flood event.* Nature Geoscience, Vol. 3, p. 477-481.

Chapter 17. How Old Is the Grand Canyon?
Page 174

A majority of geologists studying the canyon hold to the view that most of the incision occurred within the last 6 million years; a smaller group argue that incision of an earlier ("proto") canyon began long before that: Ranney, W., 2012, *Carving Grand Canyon: Evidence, Theories, and Mystery, 2nd Ed.* Grand Canyon Association, Grand Canyon, 190 pp., see p. 170.

Tributary canyons in the eastern Grand Canyon oddly point upstream, suggesting to some that the channel here originally developed with drainage to the north: Ranney, W. D., 1998, *Geomorphic evidence for the evolution of the Colorado River in the Little Colorado-Marble Canyon area, Grand Canyon, Arizona (abstract).* The Geological Society of America, Abstracts with Programs, Vol. 30, no. 6, p. 34.

Page 175

By dating certain types of speleothems, Hill and her colleagues have been able to determine when the ancient water table began to drop in elevation as the western canyon was cutting downward: Polyak, V., Hill, C. and Asmerom, Y., 2008, *Age and evolution of Grand Canyon revealed by U–Pb dating of water-table-type speleothems.* Science, Vol. 319, No. 5868, p. 1377–1380. • Hill, C. A. and Polyak, V. J., 2010, *Karst hydrology of Grand Canyon, Arizona, USA.* Journal of Hydrology, Vol. 390, p. 169-181.

Hill has argued that streams on the eastern and western sides of the Kaibab uplift were likely first connected by water flowing under the Kaibab uplift through fractures and caves, a type of stream piracy called *karst piracy:* Hill, C. A., Eberz, N. and Buecher, R. H., 2008, *A Karst Connection model for Grand Canyon, Arizona, USA.* Geomorphology, Vol. 95, p. 316–334 • Hill, C. A. and Polyak, V.J., 2014, *Karst piracy: A mechanism for integrating the Colorado River across the Kaibab uplift, Grand Canyon, Arizona, USA.* Geosphere, Vol. 10, No. 4, p. 1-14.

Page 177

Young Earth proponents do not agree on which Grand Canyon/Grand Staircase layers should be considered to be flood deposits and which should be considered to be pre- and post-flood deposits. Examples: Oard and Froede say the entire Grand Canyon Supergroup was deposited during the early flood: Oard, M. and Froede Jr., C., 2008, *Where is the Pre-Flood/Flood Boundary?*; Creation Research Society Quarterly, Vol. 43, No. 1, p. 24-39. • Austin states that only the very top formation of the Grand Canyon Supergroup was deposited during the early flood, everything below that is pre-flood: Austin and Wise, 1994, *The Pre-Flood/Flood Boundary: As Defined in Grand Canyon, Arizona and Eastern Mojave Desert, California.* In: *Proceedings of the Third International Conference on Creationism,* p. 37-47. • Oard maintains that most Cenozoic layers (65 MY to present, i.e. those higher than the Grand Staircase) were also deposited by the flood: Oard, M. J., 1996, *Where is the Flood/Post Flood Boundary in the Rock Record?* Creation ex nihilo Technical Journal, Vol. 10, No. 2. • Austin and Snelling indicate that the top of the Grand Staircase marks the end of flood deposition: Austin, 1994, *Grand Canyon: Monument to Catastrophe*, p. 58; Snelling, 2009, *Earth's Catastrophic Past: Geology, Creation and the Flood*, p. 751-761. [For review of conflicting Young Earth arguments, see: Senter, 2011, *The Defeat of Flood Geology by Flood Geology*, p. 1.1-1.14.]

Much has been learned in the last 50 years – and all of it is increasingly at odds with the flood geology model: Collins, L. G., 2015, *When Was Grand Canyon Carved— Millions of Years Ago or Thousands of Years Ago? How Do We Know?,* Reports of the National Center for Science Education, Vol.35, No. 4, p.2.1-2.8.

Chapter 18. Life in the Canyon: Packrats, Pollen, and Giant Sloths
Page 180

A number of packrat middens in Grand Canyon caves have been carbon-14 dated to be about 20,000 to 10,000 years old: Cole, K. L. and Arundel, S. T., 2005, *Carbon isotopes from fossil packrat pellets and elevational movement of Utah agave plants reveal the Younger Dryas cold period in Grand Canyon, Arizona.* Geology, Vol. 33, No. 9, p. 697-760. • Cole, K. L. and Mayer, L., 1982, *Use of packrat middens to determine rates of cliff retreat in the eastern Grand Canyon, Arizona.* Geology, Vol. 10, p. 597-599.

Packrat middens and dung in a number of caves contain organic material that reveals long periods of time when the lower reaches of the canyon were populated by juniper and ash forests: Mead, J. I. and Phillips, A. M., 1981, *The Late Pleistocene and Holocene fauna and flora of Vulture Cave, Grand Canyon, Arizona.* Southwest Naturalist, Vol. 26, p. 257-288. • Phillips, A.M., 1984, *Shasta ground sloth extinction: fossil packrat midden evidence from the western Grand Canyon.* In: Martin, P.S. and Klein, R.G. (eds.), *Quaternary Extinctions,* University of Arizona Press, p. 148-158. • Phillips, A.M. and Van Devender, T.R., 1974, *Pleistocene packrat middens from the lower Grand Canyon of Arizona.* Journal of the Arizona Academy of Science, Vol. 9, p. 117-119. • Van Devender, T. R. and Mead, J. I.,1976, *Late Pleistocene and modern plant communities of Shinumo Creek and Peach Springs Wash, Lower Grand Canyon, Arizona.* Journal of Arizona Academy of Science, Vol. 11, p. 16-22.

Page 182

Samples from the deepest dung deposits in the cave have yielded ages greater than 35,000 years, suggesting that the cave was occupied on and off by giant sloths for almost 25,000 years: Martin, P. S., Sabels, B. E. and Shutler, D., 1961, *Rampart Cave coprolite and ecology of the Shasta ground sloth.* American Journal of Science, Vol. 259, p. 102-127. • Mead, J. I., 1981, *The last 30,000 years of faunal history within the Grand Canyon, Arizona.* Quaternary Research, Vol. 15, p. 311-326. • Mead, J. I. and Agenbroad, L.D., 1992, *Isotope dating of Pleistocene dung deposits from the Colorado Plateau, Arizona and Utah.* Radiocarbon, Vol. 34, No. 1, p. 1-19.

The earliest human artifacts in Grand Canyon caves have been radiocarbon-dated to about 6,000 years ago: Mead, J. I., 1981, *The last 30,000 years of faunal history within the Grand Canyon, Arizona.* Quaternary Research, Vol. 15, p. 311-326. • Bachhuber, F.W., Rowland, S. and Huntoon, P., 1987, *Geology of the Lower Grand Canyon and Upper Lake Mead by boat – an overview.* In: Davis, G.H. and Vandenholder, E.M. (eds.), *Geologic Diversity of Arizona and its margins: Excursions to Choice Areas,* Arizona Bureau of Geology and Mineral Technology, Geological Survey Branch, Special Paper No. 5, p. 39-51.

Chapter 19. River to Rim: Putting All the Pieces Together
Page 190

There is clearly a complex history here that would require a lengthy discussion to fully describe: Timmons, Bloch, Fletcher, Karlstrom, Heizler and Crossey, 2012, *The Grand Canyon Unkar Group: Mesoproterozoic basin formation in the continental interior during supercontinent assembly.* In: *Grand Canyon Geology: Two Billion Years of Earth's History,* p. 25-47. • Dehler, C. M., Porter, S. M. and Timmons, J. M., 2012, *The Neoproterozoic Earth system revealed from the Chuar Group of Grand Canyon.* In: *Grand Canyon Geology: Two Billion Years of Earth's History,* p. 49-72.

Page 196

Of the thousands of known trilobite species, forty-seven have been found in the Tonto Group: Thayer, D., 2009, *An Introduction to Grand Canyon Fossils.* Grand Canyon Association, 64 pp., see p. 25.

Page 203

Box: Flood geologists insist that the presence of dolomite (in the Coconino Sandstone) means the whole system is marine: Whitmore, J. H., Strom, R, Cheung, S. and Garner, P. A., 2014, *The Petrology of the Coconino Sandstone (Permian), Arizona, USA.* Answers Research Journal, Vol. 7, p. 499–532.

INDEX

Bold = shown in a photo or illustration

Erosion of, 48-49
Flood geology explanation for, 34-35
Formations in, 34, **41**
Modern geology explanation for, 46-47
Sedimentary structures in, 71
Thickness, 35
Unconformity underneath, 191-192
Grand Canyon Village, 15, 18, 107, 187
Grand Staircase, 36, **37**, **39**, 41, **43**, **49**, 53, **54**, 57, 79, 127, 129, 131, 135, 136, 139-142, 145, 146, 149-150, 153, 161, 173, 177, 191
Geographic description, 32-34
Origin according to flood geology, 37-38
Origin according to modern geology, 49
Grand Wash Cliffs, 15, **33**
Gravity current, 62, **64**
Hyper-concentrated, 64
Great Unconformity, **42**, **43**, **48**, **74**, 99, 101, 102, 115, 123, 136-137, **189**, **191**, 192, 193
Gulf of California, 134, **168**
Gypsum, 56, 61, 203

H

Hakatai Shale, **47**, **84**, 91, 96, 151, **189**
Fossil pollen present?, 145, 150
Igneous intrusion dated by flood geologist, 96
Half life, 90-93
Halite, 56, 61
Headward erosion, 174, **175**
Hermit Formation, **32**, **41**, **44**, 59-60, 62, **82**, **98**, 138, 149, 154, **199**, **200**, 201
Himalaya Mountains, 113-115
Hoodoos, 38, 50, **51**
Hopi Lake (Lake Bidohochi), **163**, 164, 165
Horn coral, fossil, **139**
Hurricane Fault, **120**, **122**
Hurricane Monocline, 120, 124
Hydrodynamic sorting, 142
Hydroplate hypothesis, 36

I

Imperial Formation, **167**
Imprints, see trace fossils
Inner Gorge, 15, **19**, **30**, **44**, **45**, 46, 167, 193
Isotopes, 89
Ives Expedition, **9**

J

Jurassic Period, 32, 86, **132**, 140, **146**, 150

K

Kaibab arch or uplift, **33**, 125, **163**, 164, 166, 166-169, 173, **174**, 175-176

Kaibab Formation, **32**, 36-37, **41**, **44**, 49, **60**, 61, 79, **82**, 87, 89, 101, **122**, 132, 137, **139**, **143**, 173, 199, 202
Kaibab Monocline, **120**, **123**, **124**, **238-239**
Kaibab Suspension Bridge, **186**, **188**
Kaiparowits Formation, **33**, 132, **140**
Karst, 103, 105, 107, 175-176
Karst piracy, **175**
Kayenta Formation, **33**, 132

L

Lake Mead, **17**
Landslides (submarine), 64
Laramide Tectonic Episode, 115, 118, 121
Lava dams, 95-96
Lava flows, **91**, 92, 95, 97
Age of, 96
Least astonishment, see Occam's Razor
Lees Ferry, 15, **16**, **17**
Lepidodendron (trees), 103, **105**, **146-147**, 150
Limestone, 36, **39**, 44, 47, 49, 50, 56, 59, **60**, 61-63, **72**, 73, 76, **77**, 78-79, 84, 103, **104-107**, 111, **129**, **130**, **138**, **142**, 163, 166, **178**, 192, 193, **194**, 195, 196, **197**, 199, 201-203
Cave formation in, 77-78
Formation according to flood geology, 63, **111**, 195
Formation according to modern geology, 56, 59-61
Unlike Mt. St. Helens deposits, 73-74
Linnaean classification system, **135**
Lycopods, fossil, 147, 148, **149**, 150

M

Mammals, fossil, 35, 49-50, 131-136, 180
Manakacha Formation (in Supai Group), 61
Mancos Shale, 141
Marble Canyon, **15**, **17**, **18**, **72**, **162**
Mars Rover Curiosity, **93**
Megaflood, Channeled Scablands, 168-169
Mesozoic Era, 32, 34, 43, 49-50, **86**, 131-134, 139-140, 141, 146-147
Metamorphic rock, 34, 35, **43**, **43**, 44-48, 53, 58, **74**, 81, 92, **124**, 167
Metamorphism, 35, 42, 45, 92, 188
Mid-Atlantic Ridge, **94**, 110, 111
Mississippian Period, 86, 106, 132, 137, 146, 197-198
Moenkopi Formation, **32**, **33**, **41**, **122**, **139**
Molluscs, fossil, 139
Monoclines, 120, **123**, **124**
Kaibab Monocline, **120**, 123, **124**, **238-239**
Supai Monocline, **120**, 124
Toroweap Monocline, **120**, 124
Moenave Formation, **33**
Morris, Henry, 23, 87, 164

VULCANS THRONE

A volcano (Vulcans Throne) and lava flows, Western Grand
Canyon. *Photo by W. Kenneth Hamblin. Courtesy of the Hamblin Family.*

BIOGRAPHIES

Carol Hill has been working in the Grand Canyon for over 17 years and has published articles on the geology of the canyon in *Science, Geomorphology, Journal of Hydrology, and Geosphere.* Her specialty is caves and karst, and she is the author of *Cave Minerals of the World, Geology of Carlsbad Cavern, and Geology of the Delaware Basin.* She has been featured on NOVA and on National Geographic's *Naked Science* program. Carol is a Fellow of the American Scientific Affiliation and has written a number of articles for *Perspectives on Science and Christian Faith.*

Gregg Davidson has been a professor in the Geology and Geological Engineering Department at the University of Mississippi for the last twenty years, and has served as the department chair since 2013. He earned a B.S. in geology from Wheaton College, and a Ph. D. in hydrology from the University of Arizona, completing some of his research in the same radiocarbon lab that dated the Dead Sea Scrolls and the Shroud of Turin. He is an active writer and speaker on the subject of science and the Bible, with a book titled *When Faith and Science Collide,* and articles written for *Modern Reformation, Christian Research Journal,* and *Perspectives on Science and Christian Faith.*

Tim Helble recently retired from a hydrologist position at the National Weather Service. During part of his time in that agency, he forecast floods and water supply in the Colorado River and Great Basins. While obtaining his Masters degree at the University of Arizona, Tim held a seasonal hydrologist position at Grand Canyon National Park. Tim has been following the Young Earth Creationist movement since the late 1970s, and published an apologetics article on the Coconino Sandstone in *Perspectives on Science and Christian Faith.* He worked as a cartographer in the private sector and contributed many of the illustrations and photos in this book.

Wayne Ranney is a geologist and trail guide based in Flagstaff, Arizona. He became interested in geology while working as a backcountry ranger at Grand Canyon National Park and later received his Bachelor's and Master's degrees from Northern Arizona University. He has subsequently worked as a geologic lecturer on shipboard expeditions in places such as Antarctica, Africa, the Amazon, and the North and South Poles. He is the author or co-author of eight books including *Sedona Through Time, Ancient Landscapes of the Colorado Plateau,* and *Carving Grand Canyon,* which has sold almost 30,000 copies and is now in its 2nd edition.

Joel Duff grew up among the rocks of western Colorado and Utah. He received his B.S. in biology from Calvin College and M.S. and Ph.D. in botany from the University of Tennessee in 1995, and he is

currently a Professor of Biology at the University of Akron in Ohio. Joel has worked on numerous plant and animal systems using molecular methods to understand biological diversity. An author of more than 40 research articles, he has also published in *Perspectives on Science and Christian Faith,* is a speaker for Solid Rock Lectures (a science and faith ministry), and maintains a blog, *Naturalis Historia* where he writes about the intersection of science and faith.

David Elliott trained in Europe before joining the Geology Department at Northern Arizona University, where he has been for 35 years, the last twenty-eight as a Professor. He

is a paleontologist who has worked extensively in western North America, the Canadian Arctic, and Western Europe, and who has published more than 75 papers, a number of them on the paleontology of the Grand Canyon. He has also been the editor for the *Journal of Vertebrate*

Paleontology and a number of books and symposium volumes. He is co-author of two sections in the popular book *Grand Canyon Geology.*

Stephen O. Moshier discovered geology in high school during the Apollo explorations of the Moon and on an expedition to Philmont Scout Ranch in New Mexico. He has degrees in geology from Virginia Tech, Binghamton University and Louisiana State University, and has taught at Wheaton College since 1991. Steve's research includes the sedimentology and oil reservoir properties of limestone, and geoarchaeology in Egypt and Israel. He is a fellow of the American Scientific Affiliation, an

organization of Christians in the sciences, engineering and medicine. Steve contributes to a team-taught class at Wheaton College that introduces students to scientific and biblical perspectives on the origins of the universe, earth, and life.

Ralph Stearley is a paleontologist with broad interests in the history of life. He received his B.A. in biological anthropology from the University of Missouri and M.S. and Ph.D. in geosciences with emphases on paleontology from the University of Utah and University of Michigan, respectively. He is Professor of Geology at Calvin College, where he has taught since 1992. His published research has included work on

marine invertebrate ecology and paleoecology, fluvial taphonomy, the systematics and history of salmonid fishes, and Pleistocene mammalian biogeography. Together with Davis Young, Ralph co-authored a critique of flood geology, entitled *The Bible, Rocks and Time,* published by InterVarsity Press in 2008.

Bryan Tapp is a structural geologist specializing in deformation mechanisms in carbonate rocks, and fold and fracture mechanics. He has applied his expertise

to the analysis of fractured petroleum reservoirs, fractured aquifers, characterization of inversion systems, rock fabric formation, and fold dynamics. He has taught at The University of Tulsa for 30 years. He won the college teaching excellence award twice, and was named University Outstanding Professor in 2002. He currently serves as the Chairman of the Geosciences Department.

"One of my favorite photos of the eastern portion of the Grand Canyon." *Photo by Bryan Tapp.*

Roger Wiens is a planetary scientist and space adventurer who has worked at the University of California and Caltech, and is currently at Los Alamos National Laboratory. He was the flight payload lead for NASA's *Genesis*

mission which brought solar particles back to Earth in 2004. He is the principal investigator of the SuperCam and ChemCam laser instruments on two Mars rover missions. He has published over 125 papers on the Earth, Moon, Mars, Io, comets, and the Sun; an online book *Radiometric Dating: A Christian Perspective;* and *Red Rover,* a book about his work on the Mars Mission.

Ken Wolgemuth is a petroleum geologist with 35 years of experience in the oil industry. He received a B.S. in chemistry from Wheaton College and a Ph.D. in geochemistry from Columbia University.

He managed book publications for the American Association of Petroleum Geologists, authored papers on oil reservoirs, and for the last 16 years, has taught oil industry short courses. He specializes in teaching geology to non-geologists, founded Solid Rock Lectures to communicate the geology of the Earth to pastors, and co-authored articles for the *Christian Research Journal,* and *BioLogos.* Ken is a Fellow of the American Scientific Affiliation.

Bronze Black is a graphic designer, illustrator and photographer who lives in Flagstaff, Arizona. With a degree in geology from Northern Arizona University, he has been studying the landscape and geography of the American Southwest for over 20 years. In 1995, he traveled through Grand Canyon on the Colorado River for the first time and fell in love with the place, where  he has been a commercial river guide for the last 15 years. He has led the research and development and/or contributed to numerous books, interpretive materials, and maps on the Southwest including: *Carving Grand Canyon, Anatomy of Grand Canyon,* the *Grand Canyon Adventure Board Game,* the *Grand Canyon Map and Guide, River and Desert Plants of the Grand Canyon, Sedona Through Time,* the *Sedona Map and Guide,* and *Volcanoes of Northern Arizona.*

Susan Coman is a graphic designer, writer and photographer from Tulsa, Oklahoma. She has designed and produced more than 45 geology, geophysics, and scientific books for various organizations, including the American Association of Petroleum Geologists, Society of Exploration Geophysicists, The University of Tulsa, Society for Sedimentary Geology (SEPM), Tulsa Geological Society, and Gas Processors Association. Coman is the winner of two Oklahoma Book Awards and was presented the NCCJ's Beth Macklin Award of Excellence. In addition to  scientific projects, she has published numerous books including *A Community of Faith, Will Rogers: The Cherokee Kid, Thomas Gilcrease,* and *The Ithaca Gun Company.* After developing the magazine, *River's Edge,* in 2012, she contributed to the publishing of the book, *The Tulsa River,* which had been serialized in the magazine.

View looking southeast from Point Imperial on the North Rim. *Photo by Bronze Black.*

ZOROASTER TEMPLE   WOTANS THRONE

VISHNU TEMPLE

CAPE ROYAL

TAPEATS

SUPERGROUP

"The various authors of this book have done us all a tremendous service in their patient and clear exposition of geological thinking about the Grand Canyon (a magnificent place in its own right!). They are all clear that the "conflict" we've all heard about is not between "the Bible" and "Science," but rather between interpretations of the Bible and the sciences. Those of us who study and respect the Bible will appreciate this calm laying out of the sciences, and of their discovery of the processes that appear to have been at work. These are God's processes after all! I urge everyone to read this, believer or not — you will enjoy it."

— C. JOHN ("JACK") COLLINS
PROFESSOR OF OLD TESTAMENT, COVENANT THEOLOGICAL SEMINARY

"This book on the Grand Canyon clearly lays out the complex history of the formation of this wonder of the natural world over many millions of years. This volume is irenic in spirit, scientifically informed, and biblically sound. I find this work by well-qualified, faith-filled scientists convincing—a book desperately needed in a day when critics consider the Scriptures and science to be in serious conflict and various well-meaning Christians fail to fully account for the available geological evidence."

— PAUL COPAN
PROFESSOR AND PLEDGER FAMILY CHAIR OF PHILOSOPHY AND ETHICS,
PALM BEACH ATLANTIC UNIVERSITY

Panoramic photo from Navajo Point.
Photo by W. Kenneth Hamblin.
Courtesy of the Hamblin Family.

"*Monument to an Ancient Earth* is a richly illustrated, superbly organized, and exceptionally readable overview of the Grand Canyon and the many geologic processes that, in combination, have resulted in the myriad of features formed over Deep Time that are exposed in and around the canyon. Although this contribution by Carol Hill and Gregg Davidson and colleagues presents, in Part 1, an expertly detailed synthesis of what those who adhere to the doctrine of flood geology have used to support their predetermined answer to natural processes, the goal of the authors is certainly not to reconcile flood geology with science. Parts 2, 3, and 4 of *Monument to an Ancient Earth* offer a meaty, accurate, and sufficiently detailed overview of Earth processes. Anyone who reads this part of *Monument to an Ancient Earth* will understand how the scientific method has been used, repeatedly, to provide us with a growing understanding of natural processes and will conclude that the concept of Deep Time is inescapable. In Part 5, Chapter 20 provides an overview of the scientific evidence for the formation of all of the features of the Grand Canyon and compares this evidence with the doctrine of flood geology. The authors carefully, professionally, and convincingly show that flood geology is simply not science. They pose the question, "Does it really matter?", and end with what we unquestionably should all agree with, "Truth always matters" and, as Nadine Gordimer wrote, "Truth isn't always beauty, but the hunger for it is"! I congratulate Carol Hill and Gregg Davidson and their colleagues for producing a contribution that richly celebrates the long and fascinating natural history of Earth."

— JOHN W. GEISSMAN
EMERITUS PROFESSOR, DEPARTMENT OF EARTH AND PLANETARY SCIENCES, UNIVERSITY OF NEW MEXICO
PROFESSOR AND DEPARTMENT HEAD, DEPARTMENT OF GEOSCIENCES, UNIVERSITY OF TEXAS AT DALLAS
PAST PRESIDENT, GEOLOGICAL SOCIETY OF AMERICA; EDITOR IN CHIEF, TECTONICS

KAIBAB PLATEAU

"Can Bible-believing Christians also believe that the earth is billions of years old and that the Grand Canyon could not have been formed by Noah's Flood? Yes, insist the eleven authors of this fascinating book. On page after page, professional geologists explain that "flood geology" omits essential facts and fails to explain massive amounts of evidence in the Grand Canon itself. This important book must be carefully considered by everyone involved in the debate about the age of the earth."

— **WAYNE GRUDEM,** RESEARCH PROFESSOR OF THEOLOGY AND BIBLICAL STUDIES, PHOENIX SEMINARY

"I think it is terrific. It serves 3 purposes: (1) It is a comprehensive primer on geology as we know and practice it. I enjoyed the reviews of how we think and work, it reviewed many subjects for me, and I learned some new things that I should have known. (2) It is a fabulous defender of the geology that we know in the Grand Canyon and Grand Staircase. (3) It can be applied to case studies worldwide. Everyone on Earth should be aware of what this book contains as a science text, as well as a rationale for employing scientific methods to a basic understanding of anything natural."

JOHN WARME, PROFESSOR EMERITUS, COLORADO SCHOOL OF MINES,
VETERAN OF MORE THAN 50 TRIPS DOWN THE CANYON

Panoramic view of the East Kaibab Monocline viewed from the air. *Photo by W. Kenneth Hamblin (adapted from Hamblin, 2008, Anatomy of the Grand Canyon: Panoramas of the Canyon's Geology, pp. 50-51). Courtesy of the Hamblin Family.*

VERMILION CLIFTS

MARBLE PLATFORM

KAIBAB SURFACE

"Authored by professional geologists and a biologist with a wide range of expertise, *The Grand Canyon, Monument to an Ancient Earth* is an astounding achievement. Rightly disturbed by widely held claims that Noah's (alleged global) flood was responsible for the geology of the Grand Canyon area, the authors, most of whom are Christians, have systematically contrasted the claims of so-called "flood geologists" with the actual details of Grand Canyon geology. In doing so, they have utterly demolished the flood hypothesis with an avalanche of geologic evidence sufficient to fill a canyon! Readers are introduced to an abundance of fundamental principles of geology. The "Canyon's eleven" show how these principles are applied to the interpretation of the geological processes that produced the rocks and the canyon. The authors conclude with a concise overview of the geologic history of the Grand Canyon area. The text is lucid, lively, aimed at the geologic layperson, and supported by a wealth of excellent diagrams and stunning color photographs. This fabulous, informative, eminently readable book about one of Earth's most breathtaking wonders is a pleasure to read and is itself a fitting monument to Grand Canyon."

— DAVIS A. YOUNG, PROFESSOR OF GEOLOGY, EMERITUS, CALVIN COLLEGE

"The one who states his case first seems right, until the other comes and examines him." PROVERBS 18:17

Ages of Grand Canyon Rocks

Period	Formation	Thick-ness	Age MY ago
PERMIAN	Kaibab Formation	100-650	270
PERMIAN	Toroweap Formation	100-650	273
PERMIAN	Coconino Sandstone	60-600	275
PENNSYLVANIAN	Hermit Formation	100-900	280
PENNSYLVANIAN	Supai Group — Esplanade Sandstone	200-800	285-315
PENNSYLVANIAN	Supai Group	550-800	
MISSISSIPPIAN	Surprise Canyon Fm	0-400	323
MISSISSIPPIAN	Redwall Limestone	350-875	340
DEVONIAN	Temple Butte Formation	0-450	385
CAMBRIAN	Muav Limestone	140-830	505
CAMBRIAN	Bright Angel Shale	27-450	515
CAMBRIAN	Tapeats Sandstone	0-390	525
			541
PRECAMBRIAN	Grand Canyon Supergroup — Sixty Mile Fm		740
PRECAMBRIAN	Grand Canyon Supergroup — Chuar Group (Kwagunt, Galeros)	6800	740-770
PRECAMBRIAN	Grand Canyon Supergroup — Nankoweap Fm		900
PRECAMBRIAN	Grand Canyon Supergroup — Cardenas Basalt		1100
PRECAMBRIAN	Grand Canyon Supergroup — Unkar Group (Dox, Sninumo, Hakatai, Bass)	6940	1100-1200
PRECAMBRIAN	Precambrian Basement Rocks	?	1700-1850

Courtesy of the Hamblin Family

Q = Quaternary
N = Neogene
P℮ = Paleogene

Not present in the Grand Canyon

Standard Geologic Time Scale

MY = millions of years

ERA	Period	Epoch	Dura-tion MY	Age MY ago
CENOZOIC	Q	Holocene	0.01	0.01
CENOZOIC	Q	Pleistocene	2.6	2.6
CENOZOIC	N (Tertiary)	Pliocene	2.7	5.3
CENOZOIC	N (Tertiary)	Miocene	17.7	23
CENOZOIC	P℮ (Tertiary)	Oligocene	11	34
CENOZOIC	P℮ (Tertiary)	Eocene	22	56
CENOZOIC	P℮ (Tertiary)	Paleocene	10	66
MESOZOIC	Cretaceous		79	145
MESOZOIC	Jurassic		56	201
MESOZOIC	Triassic		51	252
PALEOZOIC	Permian		47	299
PALEOZOIC	Pennsylvanian (Carboniferous)		24	323
PALEOZOIC	Mississippian (Carboniferous)		36	359
PALEOZOIC	Devonian		60	419
PALEOZOIC	Silurian		25	444
PALEOZOIC	Ordovician		41	485
PALEOZOIC	Cambrian		56	541
PRECAMBRIAN				

EON	ERA		Dura-tion MY	Age MY ago
PHANEROZOIC	CENOZOIC		66	66
PHANEROZOIC	MESOZOIC		186	252
PHANEROZOIC	PALEOZOIC		289	541
PRECAMBRIAN — PROTEROZOIC	LATE		459	1000
PRECAMBRIAN — PROTEROZOIC	MIDDLE		600	1600
PRECAMBRIAN — PROTEROZOIC	EARLY		900	2500
PRECAMBRIAN — ARCHEAN	LATE		300	2800
PRECAMBRIAN — ARCHEAN	MIDDLE		400	3200
PRECAMBRIAN — ARCHEAN	EARLY		800	4000
PRECAMBRIAN — HADEAN	EARTH FORMS		600	4600

Graphic adapted from W. Kenneth Hamblin, 2008, *Anatomy of the Grand Canyon: Panoramas of the Canyon's Geology,* p 19.